React and React Native

Fourth Edition

Build cross-platform JavaScript applications with native power for the web, desktop, and mobile

Adam Boduch

Roy Derks

Mikhail Sakhniuk

BIRMINGHAM—MUMBAI

React and React Native

Fourth Edition

Group Product Manager: Pavan Ramchandani
Publishing Product Manager: Aaron Tanna
Senior Editor: Hayden Edwards
Content Development Editor: Rashi Dubey
Technical Editor: Joseph Aloocaran
Copy Editor: Safis Editing
Project Coordinator: Rashika Ba
Proofreader: Safis Editing
Indexer: Pratik Shirodkar
Production Designer: Ponraj Dhandapani
Marketing Coordinator: Anamika Singh

First published: March 2017
Second edition: September 2018
Third edition: April 2020
Fourth edition: May 2022

Production reference: 1270522

Published by Packt Publishing Ltd.
Livery Place
35 Livery Street
Birmingham
B3 2PB, UK.

ISBN 978-1-80323-128-0

www.packt.com

For Jason, Simon, and Kevin

– Adam Boduch

For every developer out there that needs a friend on their programming journey

– Roy Derks

For my wife, Anna, and daughter, Polina

– Mikhail Sakhniuk

Contributors

About the authors

Adam Boduch has been involved in large-scale JavaScript development for nearly 15 years. Before moving to the frontend, he worked on several large-scale cloud computing products using Python and Linux. No stranger to complexity, Adam has practical experience with real-world software systems and the scaling challenges they pose.

Thanks to the React team for providing the web with this fantastic tool.

Roy Derks is a serial start-up CTO, international speaker, and author from the Netherlands. He has been working with React, React Native, and GraphQL since 2016. You might know him from the book *React Projects – Second Edition*, which was released by Packt earlier this year. Over the last few years, he has inspired tens of thousands of developers worldwide through his talks, books, workshops, and courses.

Mikhail Sakhniuk is a software engineer who is highly proficient in JavaScript, React, and React Native. He has more than 6 years of experience in developing web and mobile applications. He has worked for start-ups, fintech companies, and product companies with more than 30 million users. Currently, Mikhail is working as a senior frontend engineer. In addition, he owns and maintains a number of open source projects. He also shares his knowledge and experience through books and articles.

I'd like to thank the entire JavaScript community for creating, maintaining, and using a thousand libraries and tools that make the web faster, bigger, and more accessible.

About the reviewers

Kirill Ezhemenskii is an experienced software engineer, frontend and mobile developer, solution architect, and CTO at a healthcare company. He's a functional programming advocate and an expert in the React Stack, GraphQL, and TypeScript. Kirill is also a React Native mentor.

Sunki Baek is an experienced frontend developer who primarily works with React Native to develop cross-platform applications. His React Native journey started in 2016, shortly after React Native was introduced—his first project was to create an e-commerce app with a chat feature. Ever since then, he has been involved in e-commerce, online-to-offline commerce, and food technology projects. Currently, he is working as a senior mobile developer in the Restaurant Live Order team at SkipTheDishes.

Andrew Baisden is a software developer who has experience working across different technical stacks. Primarily skilled as a JavaScript developer, he is also familiar with the programming languages Python and C#. He has a university degree and has spent a lot of time self-training to expand his knowledge and skillset. Technical writing and content creation are two other areas where he excels. Many of his articles have been shared across social media and are used as motivation and essential resources for other developers and aspiring developers who are trying to break into the industry. With a growing audience of over 5,000 members, Andrew continues to add value wherever he goes.

Table of Contents

3

Component Properties, State, and Context

4

Getting Started with Hooks

5

Event Handling, the React Way

6

Crafting Reusable Components

7

The React Component Life Cycle

8

Validating Component Properties

9

Handling Navigation with Routes

10

Code Splitting Using Lazy Components and Suspense

11

Server-Side React Components

12

User Interface Framework Components

13

High-Performance State Updates

Part 2 - React Native

14
Why React Native?

15
React Native under the Hood

16
Kick-Starting React Native Projects

17

Building Responsive Layouts with Flexbox

18

Navigating Between Screens

19

Rendering Item Lists

20

Showing Progress

21

Geolocation and Maps

22

Collecting User Input

23

Displaying Modal Screens

24

Responding to User Gestures

25

Using Animations

26

Controlling Image Display

27

Going Offline

28

Selecting Native UI Components Using NativeBase

Part 3 – React Architecture

29
Handling Application State

30
Why GraphQL?

31
Building a GraphQL React App

Index

Other Books You May Enjoy

Preface

Over the years, React and React Native has proven itself among JavaScript developers as a popular choice for a complete and practical guide to the React ecosystem. This fourth edition comes with the latest features, enhancements, and fixes to align with React 18, while also being compatible with React Native. It includes new chapters covering critical features and concepts in modern cross-platform app development with React.

From the basics of React to popular components such as Hooks, GraphQL, and NativeBase, this definitive guide will help you become a professional React developer in a step-by-step manner.

You'll begin by learning about the essential building blocks of React components. As you advance through the chapters, you'll work with higher-level functionalities in application development and then put your knowledge to work by developing user interface components for the web and native platforms. In the concluding chapters, you'll learn how to bring your application together with robust data architecture.

By the end of this book, you'll be able to build React applications for the web and React Native applications for multiple platforms - web, mobile, and desktop with confidence.

Who this book is for

This book is for any JavaScript developer who wants to start learning how to use React and React Native for mobile and web application development. No prior knowledge of React is required; however, working knowledge of JavaScript is necessary to be able to follow along with the content covered.

What this book covers

Chapter 1, Why React?, describes what React is and why you want to use it to build your application.

Chapter 2, Rendering with JSX, teaches the basics of JSX, the markup language used by React components.

Chapter 3, Context Properties, State, and Context, introduces the core mechanisms of passing data around your React application.

Chapter 4, *Getting Started with Hooks*, shows how React Hooks can be used to extend the behavior of components.

Chapter 5, *Event Handling, the React Way*, gives an overview of how events are handled by React components.

Chapter 6, *Crafting Reusable Components*, guides you through the process of refactoring components by example.

Chapter 7, *The React Component Life Cycle*, describes the various phases that React components go through and why it's important for React developers.

Chapter 8, *Validating Component Properties*, shows you how to ensure that React component property values are as expected.

Chapter 9, *Handling Navigations with Routes*, provides plenty of examples of how to set up routing for your React web app.

Chapter 10, *Code Splitting Using Lazy Components and Suspense*, introduces code-splitting techniques that result in smaller, more efficient applications.

Chapter 11, *Server-Side React Components*, teaches you how to use Next.js to build large-scale React applications that render content on a server and a client.

Chapter 12, *User Interface Framework Components*, gives an overview of how to get started with MUI, a React component library for building UIs.

Chapter 13, *High-Performance State Updates*, goes into depth on the new features in React 18 that allow for efficient state updates and a high-performing application.

Chapter 14, *Why React Native?*, describes what the React Native library is and the differences between native mobile development.

Chapter 15, *React Native under the Hood*, gives an overview of the architecture of React Native.

Chapter 16, *Kick-Starting React Native Projects*, teaches you how to start a new React Native project.

Chapter 17, *Building Responsive Layouts with Flexbox*, describes how to create a layout and add styles.

Chapter 18, *Navigating between Screens*, shows the approaches to switching between screens in an app.

Chapter 19, *Rendering Item Lists*, describes how to implement lists of data in an application.

Chapter 20, Showing Progress, shows you how to handle process indications and progress bars.

Chapter 21, Geolocation and Maps, guides you on how to track geolocation and add a map to an app.

Chapter 22, Collecting User Input, teaches you how to create forms.

Chapter 23, Displaying Modal Screens, teaches you how to create dialog modals.

Chapter 24, Responding to User Gestures, provides examples of how to handle user gestures.

Chapter 25, Using Animations, describes how to implement animations in an app.

Chapter 26, Controlling Image Display, gives an overview of how to render images in a React Native app.

Chapter 27, Going Offline, shows how to deal with an app when a mobile phone doesn't have an internet connection.

Chapter 28, Selecting Native UI Components Using NativeBase, teaches you how to create an application using the NativeBase UI library.

Chapter 29, Handling Application State, shows you how to handle application state for both web and mobile apps.

Chapter 30, Why GraphQL?, describes what GraphQL is and how to use it.

Chapter 31, Building a React GraphQL App, shows how to handle GraphQL in React and React Native apps.

To get the most out of this book

This book assumes you have a basic understanding of the JavaScript programming language. It also assumes that you'll be following along with the examples, which require a command-line terminal, a code editor, and a web browser.

The requirements for learning React Native are the same as for React development, but to run an app on a real device, you will need an Android or iOS smartphone. In order to run iOS apps in the simulator, you will need a Mac computer.

Software/hardware covered in the book	Operating system requirements
React	Windows, macOS, or Linux
React Native	

Each chapter has its own folder in the code repository, and each example runs independently of the others. Generally speaking, you can use `npm install` *and* `npm start` *to run each example. Check the* `README` *files in each folder for more specific instructions pertaining to each specific example.*

If you are using the digital version of this book, we advise you to type the code yourself or access the code from the book's GitHub repository (a link is available in the next section). Doing so will help you avoid any potential errors related to the copying and pasting of code.

Download the example code files

You can download the example code files for this book from GitHub at `https://github.com/PacktPublishing/React-and-React-Native-4th-Edition`. If there's an update to the code, it will be updated in the GitHub repository.

We also have other code bundles from our rich catalog of books and videos available at `https://github.com/PacktPublishing/`. Check them out!

Download the color images

We also provide a PDF file that has color images of the screenshots and diagrams used in this book. You can download it here: `https://static.packt-cdn.com/downloads/9781803231280_ColorImages.pdf`.

Conventions used

There are a number of text conventions used throughout this book.

`Code in text`: Indicates code words in text, database table names, folder names, filenames, file extensions, pathnames, dummy URLs, user input, and Twitter handles. Here is an example: "You have the actual routes declared as <Route> elements."

A block of code is set as follows:

```
export default function First() {
  return <p>Feature 1, page 1</p>;
}
```

When we wish to draw your attention to a particular part of a code block, the relevant lines or items are set in bold:

```
export default function List({ data, fetchItems, refreshItems,
isRefreshing }) {
  return (
    <FlatList
      data={data}
      renderItem={({ item }) => <Text style={styles.
item}>{item.value}</Text>}
      onEndReached={fetchItems}
      onRefresh={refreshItems}
      refreshing={isRefreshing}
    />
  );
}
```

Any command-line input or output is written as follows:

```
npm install @react-navigation/bottom-tabs @react-navigation/
drawer
```

Bold: Indicates a new term, an important word, or words that you see onscreen. For instance, words in menus or dialog boxes appear in **bold**. Here is an example: "The **Container Component** will typically contain one direct child."

> **Tips or Important Notes**
> Appear like this.

Get in touch

Feedback from our readers is always welcome.

General feedback: If you have questions about any aspect of this book, email us at `customercare@packtpub.com` and mention the book title in the subject of your message.

Errata: Although we have taken every care to ensure the accuracy of our content, mistakes do happen. If you have found a mistake in this book, we would be grateful if you would report this to us. Please visit `www.packtpub.com/support/errata` and fill in the form.

Piracy: If you come across any illegal copies of our works in any form on the internet, we would be grateful if you would provide us with the location address or website name. Please contact us at `copyright@packt.com` with a link to the material.

If you are interested in becoming an author: If there is a topic that you have expertise in and you are interested in either writing or contributing to a book, please visit `authors.packtpub.com`.

Part 1 – React

In this part, we will cover the fundamentals of React tools and concepts, applying them to build high-performance web apps.

In this part, we will cover the following chapters:

- *Chapter 1, Why React?*
- *Chapter 2, Rendering with JSX*
- *Chapter 3, Context Properties, State, and Context*
- *Chapter 4, Getting Started with Hooks*
- *Chapter 5, Event Handling, the React Way*
- *Chapter 6, Crafting Reusable Components*
- *Chapter 7, The React Component Life Cycle*
- *Chapter 8, Validating Component Properties*
- *Chapter 9, Handling Navigations with Routes*
- *Chapter 10, Code Splitting Using Lazy Components and Suspense*
- *Chapter 11, Server-Side React Components*
- *Chapter 12, User Interface Framework Components*
- *Chapter 13, High-Performance State Updates*

1
Why React?

If you're reading this book, you probably know what React is. If not, don't worry. I'll do my best to keep philosophical definitions to a minimum. However, this is a long book with a lot of content, so I feel that setting the tone is an appropriate first step. Yes, the goal is to learn React and React Native. But it's also to put together a lasting architecture that can handle everything we want to build with React today and in the future.

This chapter starts with a brief explanation of why React exists. Then, we'll think about the simplicity of React and how it is able to handle many of the typical performance issues faced by web developers. Next, we'll go over the declarative philosophy of React and the level of abstraction that React programmers can expect to work with. Finally, we'll touch on some of the major features of React.

Once you have a conceptual understanding of React and how it solves problems with UI development, you'll be better equipped to tackle the remainder of the book. This chapter will cover the following topics:

- What is React?
- React features
- What's new in React 18?

What is React?

I think the one-line description of **React** on its home page (`https://reactjs.org/`) is concise and accurate:

"A JavaScript library for building user interfaces."

It's a library for building **User Interfaces** (**UIs**). This is perfect because, as it turns out, this is all we want most of the time. I think the best part about this description is everything that it leaves out. It's not a mega framework. It's not a full-stack solution that's going to handle everything from the database to real-time updates over WebSocket connections. We might not actually want most of these prepackaged solutions. If React isn't a framework, then what is it exactly?

React is just the view layer

React is generally thought of as the view layer in an application. You might have used a library such as Handlebars or jQuery in the past. Just as jQuery manipulates UI elements and Handlebars templates are inserted into a page, React components change what the user sees. The following diagram illustrates where React fits in our frontend code:

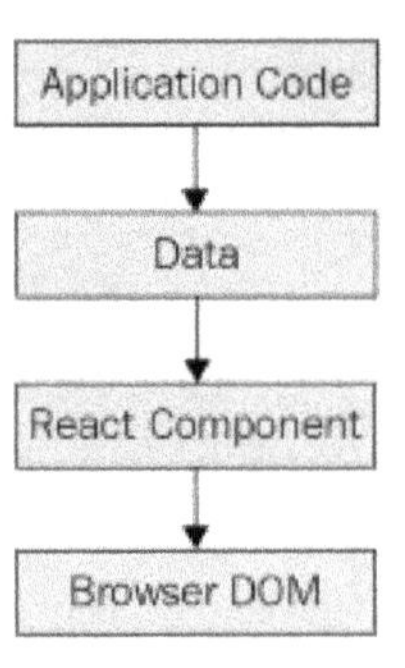

Figure 1.1 – The layers of a React application

This is all there is to React – the core concept. Of course, there will be subtle variations to this theme as we make our way through the book, but the flow is more or less the same. We have some application logic that generates some data. We want to render this data to the UI, so we pass it to a React Component, which handles the job of getting the HTML into the page.

You may wonder what the big deal is; React appears to be yet another rendering technology. We'll touch on some of the key areas where React can simplify application development in the remaining sections of the chapter.

Simplicity is good

React doesn't have many moving parts to learn about and understand. Internally, there's a lot going on, and we'll touch on these things throughout the book. The advantage of having a small API to work with is that you can spend more time familiarizing yourself with it, experimenting with it, and so on. The opposite is true of large frameworks, where all of your time is devoted to figuring out how everything works. The following diagram gives you a rough idea of the APIs that we have to think about when programming with React:

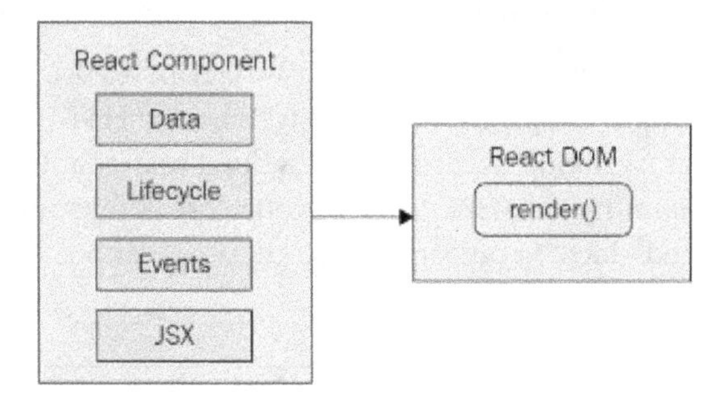

Figure 1.2 – The simplicity of the React API

React is divided into two major APIs:

- **The React Component API**: These are the parts of the page that are rendered by the React DOM.
- **React DOM**: This is the API that's used to perform the rendering on a web page.

Within a React component, we have the following areas to think about:

- **Data**: This is data that comes from somewhere (the component doesn't care where) and is rendered by the component.
- **Lifecycle**: This consists of methods or Hooks that we implement to respond to the component's entering and exiting phases of the React rendering process as they happen over time – for example, one phase of the life cycle is when the component is about to be rendered.
- **Events**: These are the code that we write for responding to user interactions.
- **JSX**: This is the syntax of React components used to describe UI structures.

Don't fixate on what these different areas of the React API represent just yet. The takeaway here is that React, by nature, is simple. Just look at how little there is to figure out! This means that we don't have to spend a ton of time going through API details here. Instead, once you pick up on the basics, we can spend more time on nuanced React usage patterns that fit in nicely with declarative UI structures.

Declarative UI structures

React newcomers have a hard time getting to grips with the idea that components mix in markup with their JavaScript in order to declare UI structures. If you've looked at React examples and had the same adverse reaction, don't worry. Initially, we're all skeptical of this approach, and I think the reason is that we've been conditioned for decades by the separation of concerns principle. This principle states that different concerns, such as logic and presentation, should be separate from one another. Now, whenever we see things mixed together, we automatically assume that this is bad and shouldn't happen.

The syntax used by React components is called JSX (JavaScript XML). A component renders content by returning some JSX. The JSX itself is usually HTML markup, mixed with custom tags for React components. The specifics don't matter at this point; we'll go into detail in the coming chapters. What's groundbreaking about the declarative JSX approach is that we don't have to perform little micro-operations to change the content of a component.

> **Important Note**
>
> Although I won't be following the convention in this book, some React developers prefer the `.jsx` extension instead of `.js` for their components.

For example, think about using something such as jQuery to build your application. You have a page with some content on it, and you want to add a class to a paragraph when a button is clicked. Performing these steps is easy enough. This is called imperative programming, and it's problematic for UI development. While this example of changing the class of an element is simple, real applications tend to involve more than three or four steps to make something happen.

React components don't require you to execute steps in an imperative way. This is why JSX is central to React components. The XML-style syntax makes it easy to describe what the UI should look like – that is, what are the HTML elements that this component is going to render? This is called declarative programming and is very well suited for UI development. Once you've declared your UI structure, you need to specify how it changes over time.

Data changes over time

Another area that's difficult for React newcomers to grasp is the idea that JSX is like a static string, representing a chunk of rendered output. This is where time and data come into play. React components rely on data being passed into them. This data represents the dynamic parts of the UI – for example, a UI element that's rendered based on a Boolean value could change the next time the component is rendered. Here's a diagram illustrating the idea:

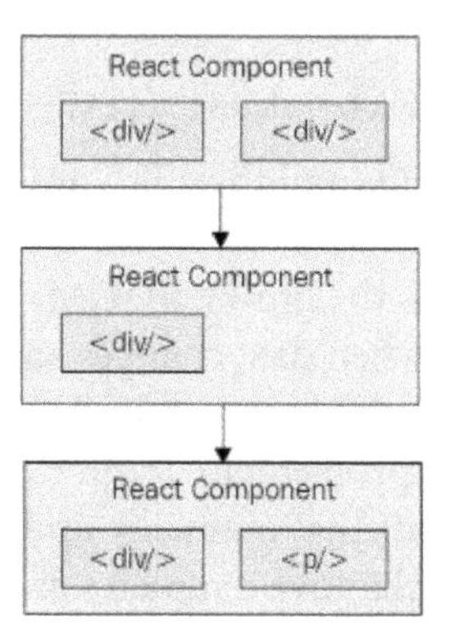

Figure 1.3 – React components changing over time

Each time the React component is rendered, it's like taking a snapshot of the JSX at that exact moment in time. As your application moves forward through time, you have an ordered collection of rendered UI components. In addition to declaratively describing what a UI should be, re-rendering the same JSX content makes things much easier for developers. The challenge is making sure that React can handle the performance demands of this approach.

Performance matters

Using React to build UIs means that we can declare the structure of the UI with JSX. This is less error-prone than the imperative approach of assembling the UI piece by piece. However, the declarative approach does present a challenge –performance.

For example, having a declarative UI structure is fine for the initial rendering because there's nothing on the page yet. So, the React renderer can look at the structure declared in JSX and render it in the DOM browser.

> **Important Note**
>
> The **Document Object Model** (**DOM**) represents HTML in the browser after it has been rendered. The DOM API is how JavaScript is able to change content on a page.

This concept is illustrated in the following diagram:

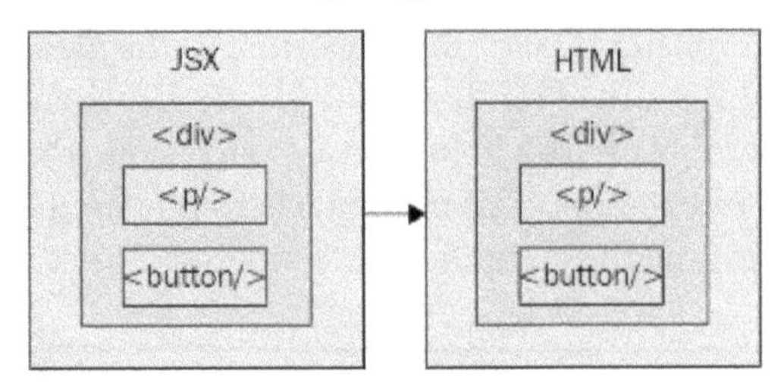

Figure 1.4 – How JSX syntax translates to HTML in the browser DOM

On the initial render, React components and their JSX are no different from other template libraries. For instance, Handlebars will render a template to HTML markup as a string, which is then inserted into the browser DOM. Where React is different from libraries such as Handlebars is when data changes and we need to re-render the component. Handlebars will just rebuild the entire HTML string, the same way it did on the initial render. Since this is problematic for performance, we often end up implementing imperative workarounds that manually update tiny bits of the DOM. We end up with a tangled mess of declarative templates and imperative code to handle the dynamic aspects of the UI.

We don't do this in React. This is what sets React apart from other view libraries. Components are declarative for the initial render, and they stay this way even as they're re-rendered. It's what React does under the hood that makes re-rendering declarative UI structures possible.

React has something called the virtual DOM, which is used to keep a representation of the real DOM elements in memory. It does this so that each time we re-render a component, it can compare the new content to the content that's already displayed on the page. Based on the difference, the virtual DOM can execute the imperative steps necessary to make the changes. So, not only do we get to keep our declarative code when we need to update the UI but React will also make sure that it's done in a performant way. Here's what this process looks like:

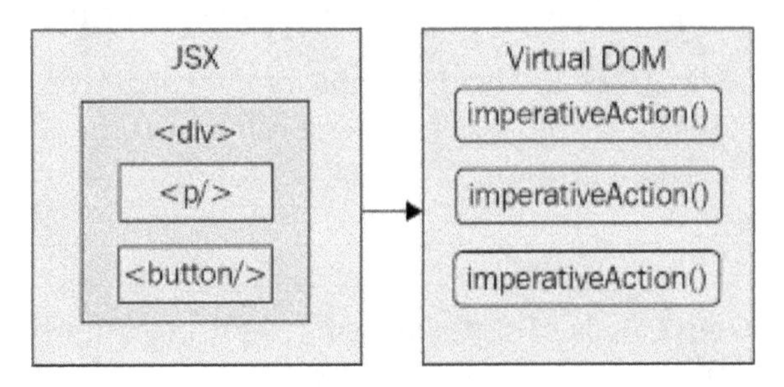

Figure 1.5 – React transpiles JSX syntax into imperative DOM API calls

> **Important Note**
>
> When you read about React, you'll often see words such as diffing and patching. Diffing means comparing old content with new content to figure out what's changed. Patching means executing the necessary DOM operations to render the new content.

As with any other JavaScript library, React is constrained by the run-to-completion nature of the main thread. For example, if the React internals are busy diffing content and patching the DOM, the browser can't respond to user input. As you'll see in the last section of this chapter, changes were made to the internal rendering algorithms in React 16 to mitigate these performance pitfalls. With performance concerns addressed, we need to make sure that we're confident that React is flexible enough to adapt to different platforms that we might want to deploy our apps to in the future.

The right level of abstraction

Another topic I want to cover at a high level before we dive into React code is abstraction.

In the preceding section, you saw how JSX syntax translates to low-level operations that update our UI. A better way to look at how React translates our declarative UI components is via the fact that we don't necessarily care what the render target is. The render target happens to be the browser DOM with React, but it isn't restricted to the browser DOM.

React has the potential to be used for any UI we want to create, on any conceivable device. We're only just starting to see this with React Native, but the possibilities are endless. I personally will not be surprised if React Toast becomes a thing, targeting toasters that can singe the rendered output of JSX onto bread. The abstraction level with React is at the right level, and it's in the right place.

The following diagram gives you an idea of how React can target more than just the browser:

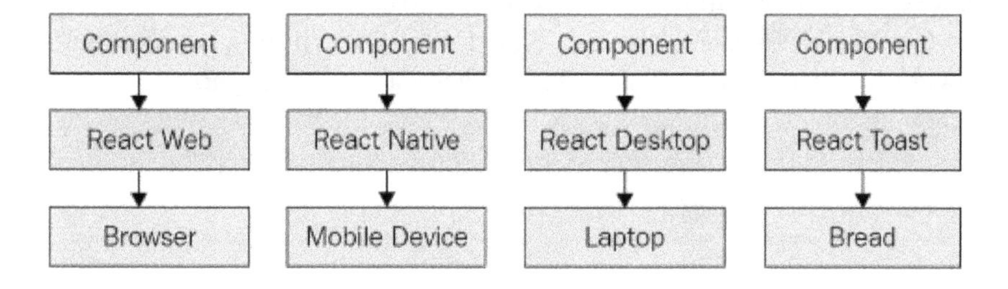

Figure 1.6 – React abstracts the target rendering environment from the components that we implement

From left to right, we have **React Web** (just plain React), **React Native**, **React Desktop**, and **React Toast**. As you can see, to target something new, the same pattern applies:

- Implement components specific to the target.
- Implement a React renderer that can perform the platform-specific operations under the hood.

This is, obviously, an oversimplification of what's actually implemented for any given React environment. But the details aren't so important to us. What's important is that we can use our React knowledge to focus on describing the structure of our UI on any platform.

> **Important Note**
> React Toast will probably never be a thing, unfortunately.

Now that you understand the role of abstractions in React, let's see what's new in React 18.

What's new in React 18?

The examples in this book are based on React 18. This release doesn't introduce sweeping API changes the way React 16 did. There are, however, two notable changes that we'll cover in more depth in *Chapter 13, High-Performance State Updates*.

Automatic batching

Batching state updates together drastically improves the performance of React applications because it reduces the number of renders to be performed. React has always had the ability to batch multiple state updates into one state update, but it was limited by where this could happen. Specifically, you could only batch state updates together inside event handler functions. The problem here is that most of our state update code runs in an asynchronous way that prevents automatic batching from happening.

React 18 removes this barrier and allows for automatic state update batching to happen anywhere. In *Chapter 13, High-Performance State Updates*, you'll see examples that compare how this worked prior to React 18 and what you can expect now.

State transitions

React 18 introduces the notion of a state transition. The idea with state transitions is that the less important state updates that take place in your application should have lower priority than state updates that should happen immediately. In *Chapter 13, High-Performance State Updates*, we'll explore the new APIs that make setting state update priority a reality in React 18.

It might not seem like much has changed in React 18, but the two major areas that we'll cover have far-reaching consequences for how React applications are implemented going forward. Existing React APIs for this version have mostly been left unchanged so that the React community can quickly adopt this latest major version upgrade without any friction.

Summary

In this chapter, you were introduced to React at a high level. React is a library, with a small API, used to build UIs. Next, you were introduced to some of the key concepts of React. We discussed the fact that React is simple because it doesn't have a lot of moving parts. Next, we looked at the declarative nature of React components and JSX. Then, you learned that React takes performance seriously and that this is how we're able to write declarative code that can be re-rendered over and over. Next, you learned about the idea of render targets and how React can easily become the UI tool of choice for all of them. Lastly, I gave you a rough overview of what's new in React 18.

That's enough introductory and conceptual stuff for now. As we make our way toward the end of the book, we'll revisit these ideas. For now, let's take a step back and nail down the basics, starting with JSX.

Further reading

Take a look at the following links for more information:

- React: `https://reactjs.org/`
- React 18: `https://reactjs.org/blog/2021/06/08/the-plan-for-react-18.html`

2
Rendering with JSX

This chapter will introduce you to JSX. JSX is the XML/HTML markup syntax that's embedded in your JavaScript code and used to declare your React components. At the lowest level, you'll use HTML markup to describe the pieces of your UI. Building React applications involves organizing these pieces of HTML markup into components. When you create a component, you add new vocabulary to JSX beyond basic HTML markup. This is where React gets interesting – when you have your own JSX tags that can use JavaScript expressions to bring your components to life. JSX is the language used to describe UIs built using React.

In this chapter, we'll cover the following:

- Your first JSX content
- Rendering HTML
- Describing the UI structure
- Creating your own JSX elements
- Using JavaScript expressions
- Fragments of JSX

Technical requirements

The code for this chapter can be found in the following directory of the accompanying GitHub repository: `https://github.com/PacktPublishing/React-and-React-Native-4th-Edition/tree/main/Chapter02`.

Your first JSX content

In this section, we'll implement the obligatory "Hello, World" JSX application. At this point, we're just dipping our toes in the water; more in-depth examples will follow. We'll also discuss what makes this syntax work well for declarative UI structures.

Hello JSX

Without further ado, here's your first JSX application:

```
import * as React from "react";
import * as ReactDOM from "react-dom";

const root =
  ReactDOM.createRoot(document.getElementById("root"));

root.render(
  <p>
    Hello, <strong>JSX</strong>
  </p>
);
```

Let's walk through what's happening here.

The `render()` function takes JSX as an argument and renders it to the DOM node passed to `ReactDOM.createRoot()`.

The actual JSX content in this example renders a paragraph with some bold text inside. There's nothing fancy going on here, so we could have just inserted this markup into the DOM directly as a plain string. However, the aim of this example is to show the basic steps involved in getting JSX rendered onto the page. Now, let's talk a little bit about the declarative UI structure.

> **Important Note**
>
> JSX is transpiled into JavaScript statements; browsers have no idea what JSX is. I would highly recommend downloading the companion code for this book from `https://github.com/PacktPublishing/React-and-React-Native-4th-Edition` and running it as you read along. Everything transpiles automatically for you; you just need to follow the simple installation steps.

Declarative UI structures

Before we move forward with more in-depth code examples, let's take a moment to reflect on our "Hello, World" example. The JSX content was short and simple. It was also declarative because it described what to render, not how to render it. Specifically, by looking at the JSX, you can see that this component will render a paragraph and some bold text within it. If this were done imperatively, there would probably be some more steps involved, and they would probably need to be performed in a specific order.

> **Important Note**
>
> I find it helpful to think of declarative as structured and imperative as ordered. It's much easier to get things right with a proper structure than to perform steps in a specific order.

The example we just implemented should give you a feel for what declarative React is all about. As we move forward in this chapter and throughout the book, the JSX markup will grow more elaborate. However, it's always going to describe what is in the UI.

The `render()` function tells React to take your JSX markup and transform it into JavaScript statements that update the UI in the most efficient way possible. This is how React enables you to declare the structure of your UI without having to think about carrying out ordered steps to update elements on the screen; an approach that often leads to bugs. Out of the box, React supports the standard HTML tags that you would find on any HTML page. Unlike static HTML, React has unique conventions that should be followed when using HTML tags.

Rendering HTML

At the end of the day, the job of a React component is to render HTML into the DOM browser. This is why JSX has support for HTML tags out of the box. In this section, we'll look at some code that renders a few of the available HTML tags. Then, we'll cover some of the conventions that are typically followed in React projects when HTML tags are used.

Built-in HTML tags

When we render JSX, element tags reference React components. Since it would be tedious to have to create components for HTML elements, React comes with HTML components. We can render any HTML tag in our JSX, and the output will be just as we'd expect.

Now, let's try rendering some of these tags:

```
import * as React from 'react';
import * as ReactDOM from 'react-dom';

const root =
  ReactDOM.createRoot(document.getElementById('root'))

root.render(
  <div>
    <button />
    <code />
    <input />
    <label />
    <p />
    <pre />
    <select />
    <table />
    <ul />
  </div>
);
```

Don't worry about the formatting of the rendered output for this example. We're making sure that we can render arbitrary HTML tags, and they render as expected, without any special definitions and imports.

> **Important Note**
>
> You may have noticed the surrounding `<div>` tag, grouping together all of the other tags as its children. This is because React needs a root element to render. Later in the chapter, you'll learn how to render adjacent elements without wrapping them in a parent element.

HTML elements rendered using JSX closely follow regular HTML element syntax with a few subtle differences regarding case-sensitivity and attributes.

HTML tag conventions

When you render HTML tags in JSX markup, the expectation is that you'll use lowercase for the tag name. In fact, capitalizing the name of an HTML tag will fail. Tag names are case-sensitive and non-HTML elements are capitalized. This way, it's easy to scan the markup and spot the built-in HTML elements versus everything else.

You can also pass HTML elements any of their standard properties. When you pass them something unexpected, a warning about the unknown property is logged. Here's an example that illustrates these ideas:

```
import * as React from "react";
import * as ReactDOM from "react-dom";

const root =
  ReactDOM.createRoot(document.getElementById("root"));

root.render(
  <button title="My Button" foo="bar">
    My Button
  </button>
);

root.render(<Button />);
```

When you run this example, it will fail to compile because React doesn't know about the `<Button>` element; it only knows about `<button>`.

> **Important Note**
> Later on in the book, I'll cover property validation for the components that you make. This avoids silent misbehavior, as seen with the `foo` property in this example.

You can use any valid HTML tags as JSX tags, as long as you remember that they're case-sensitive and that you need to pass the correct attribute names. In addition to simple HTML tags that only have attribute values, you can use HTML tags to describe the structure of your page content.

Describing UI structures

JSX is capable of describing screen elements in a way that ties them together to form a complete UI structure. Let's look at some JSX markup that declares a more elaborate structure than a single paragraph:

```
import * as React from 'react';
import * as ReactDOM from 'react-dom';

const root =
  ReactDOM.createRoot(document.getElementById('root'));

root.render(
  <section>
    <header>
      <h1>A Header</h1>
    </header>
    <nav>
      <a href="item">Nav Item</a>
    </nav>
    <main>
      <p>The main content...</p>
    </main>
    <footer>
      <small>&copy; 2021</small>
    </footer>
  </section>
);
```

This JSX markup describes a fairly sophisticated UI structure. Yet, it's easier to read than imperative code because it's XML, and XML is good for concisely expressing a hierarchical structure. This is how we want to think of our UI when it needs to change – not as an individual element or property, but the UI as a whole.

Here is what the rendered content looks like:

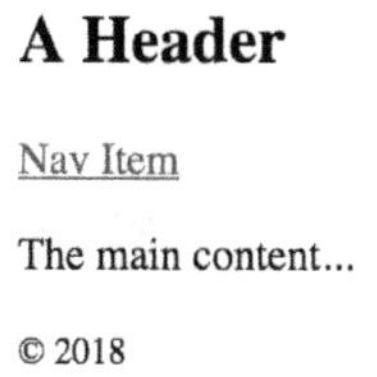

A Header

Nav Item

The main content...

© 2018

Figure 2.1 – Describing HTML tag structures using JSX syntax

There are a lot of semantic elements in this markup describing the structure of the UI. For example, the `<header>` element describes the top part of the page where the title is, and the `<main>` element describes where the main page content goes. This type of complex structure makes it clearer for developers to reason about. But before we start implementing dynamic JSX markup, let's create some of our own JSX components.

Creating your own JSX elements

Components are the fundamental building blocks of React. In fact, components are the vocabulary of JSX markup. In this section, we'll see how to encapsulate HTML markup within a component. We'll build examples that nest custom JSX elements and learn how to namespace components.

Encapsulating HTML

We create new JSX elements so that we can encapsulate larger structures. This means that instead of having to type out complex markup, you can use your custom tag. The React component returns the JSX that goes where the tag is used. Let's look at the following example:

```
import * as React from "react";
import * as ReactDOM from "react-dom";

class MyComponent extends React.Component {
  render() {
    return (
      <section>
        <h1>My Component</h1>
        <p>Content in my component...</p>
      </section>
    );
```

```
    }
}

const root =
  ReactDOM.createRoot(document.getElementById("root"));
root.render(<MyComponent />);
```

Here's what the rendered output looks like:

My Component

Content in my component...

Figure 2.2 – A component rendering encapsulated HTML markup

This is the first React component that we've implemented, so let's take a moment to dissect what's going on here. We created a class called MyComponent, which extends the Component class from React. This is how we create a new JSX element. As you can see in the call to render(), you're rendering a <MyComponent> element.

The HTML that this component encapsulates is returned by the render() method. In this case, when the JSX is rendered by react-dom, it's replaced by a <section> element and everything within it.

> **Important Note**
>
> When React renders JSX, any custom elements that you use must have their corresponding React component within the same scope. In the preceding example, the MyComponent class was declared in the same scope as the call to render(), so everything worked as expected. Usually, you'll import components, adding them to the appropriate scope. You'll see more of this as you progress through the book.

HTML elements such as <div> often take nested child elements. Let's see whether we can do the same with JSX elements, which we create by implementing components.

Nested elements

Using JSX markup is useful for describing UI structures that have parent-child relationships. Child elements are created by nesting them within another component: the parent. For example, a <li> tag is only useful as the child of a <ul> tag or a <ol> tag—you're probably going to make similar nested structures with your own React components. For this, you need to use the children property. Let's see how this works. Here's the JSX markup:

```
import * as React from "react";
import * as ReactDOM from "react-dom";

import MySection from "./MySection";
import MyButton from "./MyButton";

const root =
  ReactDOM.createRoot(document.getElementById("root"));

root.render(
  <MySection>
    <MyButton>My Button Text</MyButton>
  </MySection>
);
```

You're importing two of your own React components: `MySection` and `MyButton`.

Now, if you look at the JSX markup, you'll notice that `<MyButton>` is a child of `<MySection>`. You'll also notice that the `MyButton` component accepts text as its child, instead of more JSX elements.

Let's see how these components work, starting with `MySection`:

```
import * as React from "react";

class MySection extends React.Component {
  render() {
    return (
      <section>
        <h2>My Section</h2>
        {this.props.children}
      </section>
    );
  }
}

export default MySection;
```

This component renders a standard `<section>` HTML element, a heading, and then `{this.props.children}`. It's this last piece that allows components to access nested elements or text, and to render them.

> **Important Note**
>
> The two braces used in the preceding example are used for JavaScript expressions. I'll touch on more details of the JavaScript expression syntax found in JSX markup in the following section.

Now, let's look at the `MyButton` component:

```
import * as React from "react";

class MyButton extends React.Component {
  render() {
    return <button>{this.props.children}</button>;
  }
}

export default MyButton;
```

This component uses the exact same pattern as `MySection`; it takes the `{this.props.children}` value and surrounds it with markup. React handles the details for you. In this example, the button text is a child of `MyButton`, which is, in turn, a child of `MySection`. However, the button text is transparently passed through `MySection`. In other words, we didn't have to write any code in `MySection` to make sure that `MyButton` got its text. Pretty cool, right? Here's what the rendered output looks like:

My Section

My Button Text

Figure 2.3 – A button element rendered using child JSX values

We can further organize our components by placing them within a namespace.

Namespaced components

The custom elements that you've created so far have used simple names. A namespace provides an organizational unit for your components so that related components can share the same namespace prefix. Instead of writing `<MyComponent>` in your JSX markup, you would write `<MyNamespace.MyComponent>`. This makes it clear that `MyComponent` is part of `MyNamespace`.

Typically, `MyNamespace` would also be a component. The idea of namespacing is to have a namespace component render its child components using the namespace syntax. Let's take a look at an example:

```
import * as React from "react";
import * as ReactDOM from "react-dom";
import MyComponent from "./MyComponent";

const root =
  ReactDOM.createRoot(document.getElementById("root"));

root.render(
  <MyComponent>
    <MyComponent.First />
    <MyComponent.Second />
  </MyComponent>
);
```

This markup renders a `<MyComponent>` element with two children. Instead of writing `<First>`, we write `<MyComponent.First>`, and the same with `<MyComponent.Second>`. We want to explicitly show that `First` and `Second` belong to `MyComponent` within the markup.

Now, let's take a look at the `MyComponent` module:

```
import * as React from "react";

class First extends React.Component {
  render() {
    return <p>First...</p>;
  }
}
```

```
class Second extends React.Component {
  render() {
    return <p>Second...</p>;
  }
}

class MyComponent extends React.Component {
  render() {
    return <section>{this.props.children}</section>;
  }
}

MyComponent.First = First;
MyComponent.Second = Second;

export default MyComponent;
export { First, Second };
```

This module declares `MyComponent` as well as the other components that fall under this namespace (`First` and `Second`). It assigns the components to the namespace component (`MyComponent`) as class properties. There are a number of things that you could change in this module. For example, you don't have to directly export `First` and `Second` since they're accessible through `MyComponent`. You also don't need to define everything in the same module; you could import `First` and `Second` and assign them as class properties. Using namespaces is completely optional, and, if you use them, you should use them consistently.

You now know how to build your own React components that introduce new JSX tags in your markup. The components that we've looked at so far in this chapter have been static. That is, once we rendered them, they were never updated. JavaScript expressions are the dynamic pieces of JSX and are what cause React to update components.

Using JavaScript expressions

As you saw in the preceding section, JSX has a special syntax that allows you to embed JavaScript expressions. Any time React renders JSX content, expressions in the markup are evaluated. This is the dynamic aspect of JSX, and in this section, you'll learn how to use expressions to set property values and element text content. You'll also learn how to map collections of data to JSX elements.

Dynamic property values and text

Some HTML property or text values are static, meaning that they don't change as JSX markup is re-rendered. Other values, the values of properties or text, are based on data that is found elsewhere in the application. Remember, React is just the view layer. Let's look at an example so that you can get a feel for what the JavaScript expression syntax looks like in JSX markup:

```
import * as React from 'react';
import * as ReactDOM from 'react-dom';

const enabled = false;
const text = 'A Button';
const placeholder = 'input value...';
const size = 50;

const root =
  ReactDOM.createRoot(document.getElementById('root'))

root.render(
  <section>
    <button disabled={!enabled}>{text}</button>
    <input placeholder={placeholder} size={size} />
  </section>
);
```

Anything that is a valid JavaScript expression, including nested JSX, can go in between the braces: `{}`. For properties and text, this is often a variable name or object property. Notice, in this example, that the `!enabled` expression computes a Boolean value. Here's what the rendered output looks like:

Figure 2.4 – Dynamically changing the property value of a button

> **Important Note**
> If you're following along with the downloadable companion code, which I strongly recommend doing, try playing with these values and seeing how the rendered HTML changes.

Primitive JavaScript values are straightforward to use in JSX syntax. But what if you have an object or array that you need to transform into JSX elements?

Mapping collections to elements

Sometimes, you need to write JavaScript expressions that change the structure of your markup. In the preceding section, you learned how to use JavaScript expression syntax to dynamically change the property values of JSX elements. What about when you need to add or remove elements based on JavaScript collections?

> **Important Note**
> Throughout the book, when I refer to a JavaScript collection, I'm referring to both plain objects and arrays. Or, more generally, anything that's iterable.

```
The best way to dynamically control JSX elements is to map them
from a collection.
Let's look at an example of how this is done:
import * as React from "react";
import * as ReactDOM from "react-dom";

const array = ["First", "Second", "Third"];

const object = {
  first: 1,
  second: 2,
  third: 3,
};

const root =
  ReactDOM.createRoot(document.getElementById("root"));

root.render(
  <section>
    <h1>Array</h1>
    <ul>
      {array.map((i) => (
        <li key={i}>{i}</li>
      ))}
```

```
      </ul>

      <h1>Object</h1>
      <ul>
        {Object.keys(object).map((i) => (
          <li key={i}>
            <strong>{i}: </strong>
            {object[i]}
          </li>
        ))}
      </ul>
    </section>
  );
```

The first collection is an array called `array`, populated with string values. Moving down to the JSX markup, you can see the call to `array.map()`, which returns a new array. The mapping function is actually returning a JSX element (`<li>`), meaning that each item in the array is now represented in the markup.

> **Important Note**
>
> The result of evaluating this expression is an array. Don't worry – JSX knows how to render arrays of elements.

The object collection uses the same technique, except you have to call `Object.keys()` and then map this array. What's nice about mapping collections to JSX elements on the page is that you can control the structure of React components based on the collected data. This means that you don't have to rely on imperative logic to control the UI.

Here's what the rendered output looks like:

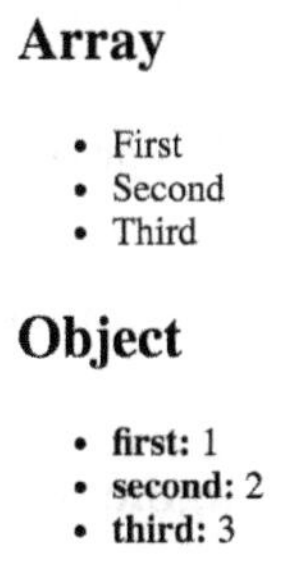

Array

- First
- Second
- Third

Object

- **first:** 1
- **second:** 2
- **third:** 3

Figure 2.5 – The result of mapping JavaScript collections to HTML elements

JavaScript expressions bring JSX content to life. React evaluates expressions and updates the HTML content based on what has already been rendered and what has changed. Understanding how to utilize these expressions is important because they're one of the most common day-to-day activities of any React developer. Now it's time to learn how to group together JSX markup without relying on HTML tags to do so.

Building fragments of JSX

React 16 introduced the concept of JSX fragments. Fragments are a way to group together chunks of markup without having to add unnecessary structure to your page. For example, a common approach is to have a React component return content wrapped in a <div> element. This element serves no real purpose and adds clutter to the DOM.

Let's look at an example. Here are two versions of a component. One uses a wrapper element, and one uses the new fragment feature:

```
import * as React from "react";
import * as ReactDOM from "react-dom";

import WithoutFragments from "./WithoutFragments";
import WithFragments from "./WithFragments";

const root =
  ReactDOM.createRoot(document.getElementById("root"));

root.render(
  <div>
    <WithoutFragments />
    <WithFragments />
  </div>
);
```

The two elements rendered are <WithoutFragments> and <WithFragments>. Here's what they look like when rendered:

Without Fragments

Adds an extra div element.

With Fragments

Doesn't have any unused DOM elements.

Figure 2.6 – Fragments help render fewer HTML tags without any visual difference

Let's compare the two approaches now.

Using wrapper elements

The first approach is to wrap sibling elements in <div>. Here's what the source looks like:

```
import * as React from "react";

class WithoutFragments extends React.Component {
  render() {
    return (
      <div>
        <h1>Without Fragments</h1>
        <p>
          Adds an extra <code>div</code> element.
        </p>
      </div>
    );
  }
}

export default WithoutFragments;
```

The essence of this component is the <h1> and <p> tags. Yet, in order to return them from render(), you have to wrap them with <div>. Indeed, inspecting the DOM using your browser dev tools reveals that <div> does nothing but add another level of structure:

```
▼<div>
   <h1>Without Fragments</h1>
  ▼<p>
     "Adds an extra "
     <code>div</code>
     " element."
   </p>
 </div>
```

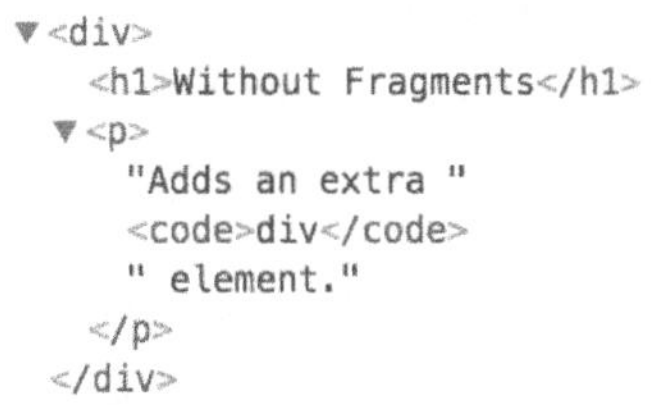

Figure 2.7 – Another level of structure in the DOM

Now, imagine an app with lots of these components—that's a lot of pointless elements! Let's see how to use fragments to avoid unnecessary tags.

Using fragments

Let's take a look at the `WithFragments` component, where we have avoided using unnecessary tags:

```
import * as React from "react";

class WithFragments extends React.Component {
  render() {
    return (
      <>
        <h1>With Fragments</h1>
        <p>Doesn't have any unused DOM elements.</p>
      </>
    );
  }
}

export default WithFragments;
```

Instead of wrapping the component content in `<div>`, the `<>` element is used. This is a special type of element that indicates that only its children need to be rendered. You can see the difference compared to the `WithoutFragments` component if you inspect the DOM:

```
<h1>With Fragments</h1>
<p>Doesn't have any unused DOM elements.</p>
```

Figure 2.8 – Less HTML in the fragment

With the advent of fragments in JSX markup, we have less HTML rendered on the page because we don't have to use tags such as `<div>` for the sole purpose of grouping elements together. Instead, when a component renders a fragment, React knows to render the fragment's child element wherever the component is used.

So, fragments enable React components to render only the essential elements; no more elements that serve no purpose will appear on the rendered page.

Summary

In this chapter, you learned about the basics of JSX, including its declarative structure, which leads to more maintainable code. Then, you wrote some code to render some basic HTML and learned about describing complex structures using JSX; every React application has at least some structure.

Next, you spent some time learning about extending the vocabulary of JSX markup by implementing your own React components, which is how you design your UI as a series of smaller pieces and glue them together to form the whole. Then, you learned how to bring dynamic content into JSX element properties, and how to map JavaScript collections to JSX elements, eliminating the need for imperative logic to control the UI display. Finally, you learned how to render fragments of JSX content using the new React 16 functionality, which prevents unnecessary HTML elements from being used.

Now that you have a feel for what it's like to render UIs by embedding declarative XML in your JavaScript modules, it's time to move on to the next chapter, where we'll take a deeper look at component properties and state.

Further reading

Refer to the following links for more information:

- Introducing JSX: `https://reactjs.org/docs/introducing-jsx.html`
- Fragments: `https://reactjs.org/docs/fragments.html`

3
Component Properties, State, and Context

React components rely on **JavaScript XML** (**JSX**) syntax, which is used to describe the structure of the UI. JSX will only get you so far—you need data to fill in the structure of your React components. The focus of this chapter is on component data, which comes in two main varieties: **properties** and **state**. Another option for passing data to components is via a context.

I'll start things off by defining what is meant by properties and state. Then, I'll walk through some examples that show you the mechanics of setting component state and passing component properties. Toward the end of this chapter, we'll build on your newfound knowledge of properties and state and introduce functional components and the container pattern. Finally, you'll learn about context and when it makes a better choice than a property for passing data to components.

In this chapter, we'll cover the following topics:

- What is component state?
- What are component properties?
- Setting a component state
- Passing property values
- Stateless components
- Container components
- Providing and consuming context

Technical requirements

The code for this chapter can be found here: `https://github.com/PacktPublishing/React-and-React-Native-4th-Edition/tree/main/Chapter03`.

What is component state?

React components declare the structure of UI elements using JSX. However, components need data if they are to be useful. For example, your component JSX might declare a `<ul>` element that maps a JavaScript collection to `<li>` elements. Where does this collection come from?

State is the dynamic part of a React component. You can declare the initial state of a component, which changes over time.

Imagine that you're rendering a component where a piece of its state is initialized to an empty array. Later on, this array is populated with data using `setState()`. This is called a **change in state**, and whenever you tell a React component to change its state, the component will automatically re-render itself, calling `render()`. The process is visualized here:

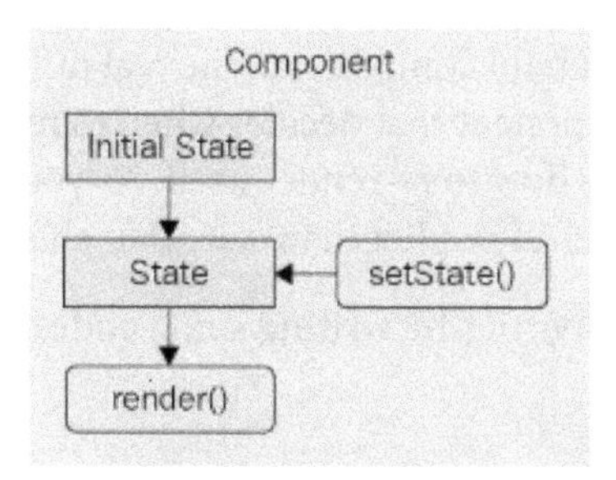

Figure 3.1 – The component state lifecycle

The state of a component is something that either the component itself can set, or other pieces of code can set, outside of the component. Now we'll look at component properties and explain how they differ from component state.

What are component properties?

Properties are used to pass data into your React components. Instead of calling a method with a new state as the argument, properties are passed only when the component is rendered, that is, you pass property values to JSX elements.

> **Note**
>
> In the context of JSX, properties are called **attributes**, probably because that's what they're called in XML parlance. In this book, properties and attributes are synonymous with one another.

Properties are different than state because they don't change after the initial render of the component. If a property value has changed, and you want to re-render the component, then we have to re-render the JSX that was used to render it in the first place. The React internals make sure that this is done efficiently. Here's a diagram of rendering and re-rendering a component using properties:

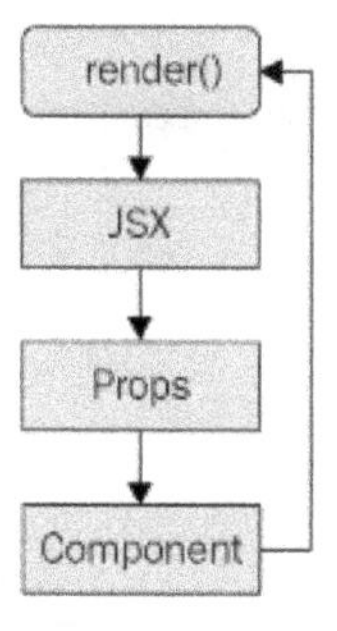

Figure 3.2 – The cycle of rendering components as properties

This looks a lot different than a stateful component. The real difference is that with properties, it's often a parent component that decides when to render the JSX. The component doesn't actually know how to re-render itself. As you'll see throughout this book, this type of top-down flow is easier to predict than state that changes all over the place.

Let's make sense of state and properties by writing some code, starting with setting the state of your components.

Setting component state

In this section, you're going to write some React code that sets the state of components. First, you'll learn about the initial state—that is, the default state of a component. Next, you'll learn how to change the state of a component, causing it to re-render itself. Finally, you'll see how a new state is merged with an existing state.

Setting initial component state

The initial state of a component isn't actually required, but if your component uses state, it should be set. This is because if the component expects certain state properties to be there and they aren't, then the component will either fail or render something unexpected. Thankfully, it's easy to set the initial component state.

The initial state of a component should always be an object with one or more properties. For example, you might have a component that uses a single array as its state. This is fine, but just make sure that you set the initial array as a property of the state object. Don't use an array as the state. The reason for this is simple: *consistency*. Every React component uses a plain object as its state.

Let's turn our attention to some code now. Here's a component that sets an initial state object:

```
import * as React from "react";

class MyComponent extends React.Component {
  state = {
    first: false,
    second: true,
  };

  render() {
    const { first, second } = this.state;
```

```
    return (
      <main>
        <section>
          <button disabled={first}>First</button>
        </section>
        <section>
          <button disabled={second}>Second</button>
        </section>
      </main>
    );
  }
}

export default MyComponent;
```

If you look at the JSX that's returned by `render()`, you can actually see the state values that this component depends on—`first` and `second`. Since you've set these properties up in the initial state, you're safe to render the component, and there won't be any surprises. For example, you could render this component only once, and it would render as expected thanks to the initial state set in `MyComponent` in the preceding code listing:

```
import * as React from "react";
import * as ReactDOM from "react-dom";
import MyComponent from "./MyComponent";

const root =
  ReactDOM.createRoot(document.getElementById("root"));
root.render(<MyComponent />);
```

Here's what the rendered output looks like:

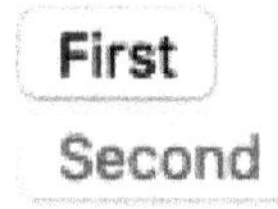

Figure 3.3 – Rendering two buttons

Setting the initial state isn't very exciting, but it's important nonetheless. Let's make the component re-render itself when the state is changed.

Creating component state

Let's create a component that has some initial state. You'll then render this component and update its state. This means that the component will be rendered twice. Let's take a look at the component:

```
import * as React from "react";

class MyComponent extends React.Component {
  state = {
    heading: "React Awesomesauce (Busy)",
    content: "Loading...",
  };

  constructor() {
    super();

    setTimeout(() => {
      this.setState({
        heading: "React Awesomesauce",
        content: "Done!",
      });
    }, 3000);
  }

  render() {
    const { heading, content } = this.state;

    return (
      <main>
        <h1>{heading}</h1>
        <p>{content}</p>
      </main>
```

```
        );
    }
}

export default MyComponent;
```

The JSX of this component depends on two state values—`heading` and `content`. The component also sets the initial values of these two state values, which means that it can be rendered without any unexpected *gotchas*. The component is first rendered with its default state. However, the interesting spot in this code is the `setTimeout()` call. After three seconds, it uses `setState()` to change the two state values. Sure enough, this change is reflected in the UI. Here's what the initial state looks like when rendered:

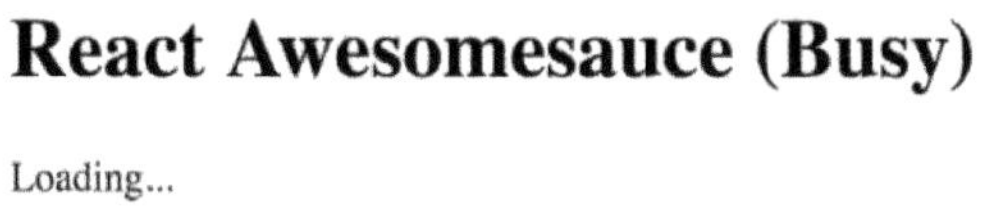

React Awesomesauce (Busy)

Loading...

Figure 3.4 –The UI while loading

Here's what the rendered output looks like after the state change:

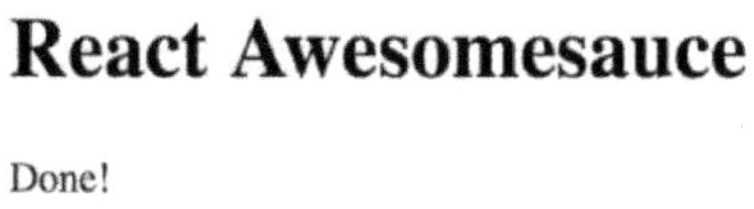

React Awesomesauce

Done!

Figure 3.5 – The UI when loading is complete

This example highlights the power of having declarative JSX syntax to describe the structure of the UI component. You declare it once and update the state of the component over time to reflect changes in the application as they happen. All the DOM interactions are optimized and hidden from view.

In this example, you replaced the entire component state—that is, the call to `setState()` passed in the same object properties found in the initial state. But what if you only want to update part of the component state?

Merging component state

When you set the state of a React component, you're actually merging the state of the component with the object that you pass to `setState()`. This is useful because it means that you can set part of the component state while leaving the rest of the state as it is. Let's look at an example now. First, let's implement a component that has some initial state set on it:

```
class MyComponent extends React.Component {
  state = {
    first: "loading...",
    second: "loading...",
    third: "loading...",
    fourth: "loading...",
    doneMessage: "finished!",
  };

  constructor() {
    super();

    setTimeout(() => {
      this.setState({ first: "done!" });
    }, 1000);

    setTimeout(() => {
      this.setState({ second: "done!" });
    }, 2000);

    setTimeout(() => {
      this.setState({ third: "done!" });
    }, 3000);

    setTimeout(() => {
      this.setState((state) => ({
        ...state,
        fourth: state.doneMessage,
      }));
    }, 4000);
```

```
  }

  render() {
    return (
      <ul>
        {Object.keys(this.state)
          .filter((key) => key !== "doneMessage")
          .map((key) => (
            <li key={key}>
              <strong>{key}: </strong>
              {this.state[key]}
            </li>
          ))}
      </ul>
    );
  }
```

This component renders the keys and values of its state—except for `doneMessage`. Each value defaults to `loading`. To iterate over objects, we have to use `Object.keys()`, which returns an array of the object keys. Next, `filter()` is used to return a new array of object keys but without the `doneMessage` value. Finally, we can call `map()` to map each object key to an `<li>` element. The value that corresponds to the key is looked up on the state object, like so: `state[key]`.

The takeaway from this example is that you can set individual state properties on components. It will efficiently re-render itself. Here's what the rendered output looks like for the initial component state:

- **first:** loading...
- **second:** loading...
- **third:** loading...
- **fourth:** loading...

Figure 3.6 – The UI while data is loading

Here's what the output looks like after three of the `setTimeout()` callbacks have run:

- **first:** done!
- **second:** done!
- **third:** done!
- **fourth:** finished!

Figure 3.7 – The UI when all async operations are complete

The fourth call to `setState()` looks different from the first three. Instead of passing a new object to merge into the existing state, you can pass a function. This function takes a state argument—the current state of the component. This is useful when you need to base state changes on current state values. In this example, the `doneMessage` value is used to set the value of `fourth`. The function then returns the new state of the component. It's up to you to merge existing state values into the new state. You can use the spread operator to do this (`...state`).

Components with state usually have an initial state. You can then change the initial values by calling `setState()`. If you only need to change part of the state, you can pass an object with only the values that you want to change and React will take care of merging the values into the overall state of the component.

Now that we've looked at the state of a component that changes over time, it's time to learn about properties that never change.

Passing property values

Properties are like state data that gets passed into components. However, properties are different from state in that they're only set once, which is when the component is rendered. In this section, you'll learn about default property values. Then, we'll look at setting property values. After this section, you should be able to grasp the differences between component state and properties.

Default property values

Default property values work a little differently than default state values. They're set as a class attribute called `defaultProps`. Let's take a look at a component that declares default property values:

```
import * as React from "react";

class MyButton extends React.Component {
```

```
  static defaultProps = {
    disabled: false,
    text: "My Button",
  };

  render() {
    const { disabled, text } = this.props;

    return <button disabled={disabled}>{text}</button>;
  }
}

export default MyButton;
```

Why not just set the default property values as an instance property, like you would with default state? The reason is that properties are immutable, and there's no need for them to be kept as an instance property value. State, on the other hand, changes all the time, so the component needs an instance-level reference to it. You can see that this component sets default property values for `disabled` and `text`. These values are only used if they're not passed in through the JSX markup used to render the component.

Let's go ahead and render this component without any properties, to make sure that the `defaultProps` values are used:

```
import * as React from "react";
import * as ReactDOM from "react-dom";
import MyButton from "./MyButton";

const root =
  ReactDOM.createRoot(document.getElementById("root"));
root.render(<MyButton />);
```

The same principle of always having a default state applies to properties too. We want to be able to render components without having to know in advance what the dynamic values of the component are. In this example, the `MyButton` component renders a `<button>` element using the default `disabled` and `text` property values.

Now, let's write some code that passes new property values to components that will override any default value for a given property.

Setting property values

React component properties are set by passing JSX attributes to the component when it is rendered. In *Chapter 8, Validating Component Properties*, I'll go into more detail about how to validate the property values that are passed to components. Now let's create a couple of components that expect different types of property values:

```
import * as React from "react";

class MyButton extends React.Component {
  render() {
    const { disabled, text } = this.props;
    return <button disabled={disabled}>{text}</button>;
  }
}

export default MyButton;
```

This simple button component expects a Boolean `disabled` property and a string `text` property. Let's create one more component that expects an array property value:

```
import * as React from "react";

class MyList extends React.Component {
  render() {
    const { items } = this.props;

    return (
      <ul>
        {items.map((i) => (
          <li key={i}>{i}</li>
        ))}
      </ul>
    );
  }
}

export default MyList;
```

You can pass just about anything you want as a property value via JSX, just as long as it's a valid JavaScript expression. The `MyList` component accepts an `items` property, an array that is mapped to `<li>` elements.

Now, let's write some code to set these property values:

```
import * as React from "react";
import * as ReactDOM from "react-dom";
import MyButton from "./MyButton";
import MyList from "./MyList";

const root =
  ReactDOM.createRoot(document.getElementById("root"));

const appState = {
  text: "My Button",
  disabled: true,
  items: ["First", "Second", "Third"],
};

function render(props) {
  root.render(
    <main>
      <MyButton text={props.text}
        disabled={props.disabled} />
      <MyList items={props.items} />
    </main>
  );
}

render(appState);

setTimeout(() => {
  appState.disabled = false;
  appState.items.push("Fourth");

  render(appState);
}, 1000);
```

The `render()` function looks like it's creating new React component instances every time it's called. React is smart enough to figure out that these components already exist, and that it only needs to figure out what the difference in output will be with the new property values. In this example, the call to `setTimeout()` causes a delay of 1 second. Then, the `appState.disabled` value is changed to `false` and the `appState.items` array has a new value added to the end of it. The call to `render()` will re-render the components with new property values.

Another takeaway from this example is that you have an `appState` object that holds on to the state of the application. Pieces of this state are then passed into components as properties when the components are rendered. State has to live somewhere and, in this case, it's outside of the component. I'll build on this topic in the next section, where you will learn how to implement stateless functional components.

Stateless components

The components you've seen so far in this book have been classes that extend the base `Component` class. It's time to learn about functional components in React. In this section, you'll learn what a functional component is by implementing one. Then, you'll learn how to set default property values for stateless functional components.

Pure functional components

A **functional React component** is just what its name suggests—a function. Picture the `render()` method of any React component that you've seen. This method, in essence, is the component. The job of a functional React component is to return JSX, just like a class-based React component. The difference is that this is all a functional component can do. It has no state and no lifecycle methods.

Why would you want to use functional components? It's a matter of simplicity more than anything else. If your component renders some JSX and does nothing else, then why bother with a class when a function is simpler?

A **pure function** is a function without side effects. That is to say, called with a given set of arguments, the function always produces the same output. This is relevant for React components because, given a set of properties, it's easier to predict what the rendered content will be. Functions that always return the same value with given argument values are easier to test as well.

Let's look at a functional component now:

```
import * as React from "react";

export default ({ disabled, text }) => (
  <button disabled={disabled}>{text}</button>
);
```

Concise, isn't it? This function returns a `<button>` element, using the properties passed in as arguments (instead of accessing them through `this.props`). This function is pure because the same content is rendered if the same `disabled` and `text` property values are passed.

Now, let's see how to render this component:

```
import * as React from "react";
import * as ReactDOM from "react-dom";
import MyButton from "./MyButton";

const root =
  ReactDOM.createRoot(document.getElementById("root"));

function render({ first, second }) {
  root.render(
    <main>
      <MyButton text={first.text}
        disabled={first.disabled} />
      <MyButton text={second.text}
        disabled={second.disabled} />
    </main>
  );
}

render({
  first: {
    text: "First Button",
    disabled: false,
  },
```

```
    second: {
      text: "Second Button",
      disabled: true,
    },
  });
```

There's zero difference between the class-based and function-based React components from a JSX point of view. The JSX looks exactly the same whether the component was declared using the class or function syntax.

The convention is to use the arrow function syntax to declare functional React components. However, it's perfectly valid to declare them using a traditional JavaScript function syntax if that's better suited to your style.

Here's what the rendered HTML looks like:

Figure 3.8 – Two buttons, one of which is disabled

Functional components rely on property values being passed to them for anything dynamic. For example, if a component renders a functional component, it usually passes in property values, and these values can change each time it is rendered. But what about default property values for functional components?

Defaults in functional components

Functional components are lightweight; they don't have any state or lifecycle. They do, however, support some metadata options. For example, you can specify the default property values of functional components the same way you would with a class component.

Here's an example of what this looks like:

```
import * as React from "react";

const MyButton = ({ disabled, text }) => (
  <button disabled={disabled}>{text}</button>
```

```
);

MyButton.defaultProps = {
  text: "My Button",
  disabled: false,
};

export default MyButton;
```

The `defaultProps` property is defined on a function instead of a class. When React encounters a functional component with this property, it knows to pass in the default properties if they're not provided via JSX.

Functional components are an important part of React applications because they're highly focused on taking property values and rendering markup that uses these values. The term *pure function* is used to indicate that a function, in our case, a React component, doesn't have any side effects. As long as you give it the same property values, the same output is rendered. Functional components can also have default property values, just as their class-based counterparts can.

You might have noticed a pattern at this point: some components have state that changes over time. These components then pass state values to other components as properties. These stateful components are called **container components**.

Container components

In this section, you're going to learn about the concept of container components. This is a common React pattern, and it brings together many of the concepts that you've learned about state and properties.

The basic premise of container components is simple: don't couple data fetching with the component that renders the data. The container is responsible for fetching the data and passing it to its child component. It contains the component responsible for rendering the data.

The idea is that you should be able to achieve some level of substitutability with this pattern. For example, a container could substitute its child component. Or a child component could be used in a different container.

Let's look at the container pattern in action, starting with the container itself:

```
import * as React from "react";
import MyList from "./MyList";

function fetchData() {
  return new Promise((resolve) => {
    setTimeout(() => {
      resolve(["First", "Second", "Third"]);
    }, 2000);
  });
}

class MyContainer extends React.Component {
  state = { items: [] };

  componentDidMount() {
    fetchData().then((items) => this.setState({ items }));
  }

  render() {
    return <MyList {...this.state} />;
  }
}

export default MyContainer;
```

The job of this component is to fetch data and to set its state. Any time the state is set, `render()` is called. This is where the child component comes in. The state of the container is passed to the `MyList` component as properties.

Let's take a look at the `MyList` component next:

```
import * as React from "react";

export default ({ items }) => (
  <ul>
    {items.map((i) => (
```

```
      <li key={i}>{i}</li>
    ))}
  </ul>
);
```

`MyList` is a functional component that expects an `items` property. Let's see how the container component is actually used:

```
import * as React from "react";
import * as ReactDOM from "react-dom";
import MyContainer from "./MyContainer";

const root =
  ReactDOM.createRoot(document.getElementById("root"));
root.render(<MyContainer />);
```

Container component design will be covered in more depth in *Chapter 6, Crafting Reusable Components*. The idea of this example is to give you a feel for the interplay between state and properties in React components.

When you load the page, you'll see the following content rendered after the three seconds it takes to simulate an HTTP request:

- First
- Second
- Third

Figure 3.9 – The rendered list of text

Containers are an important concept in React applications, as they help to separate the work of getting data and using data to render markup. You'll encounter many variations of this pattern in any given React code base. The basic idea is that the container does the work to get the data, and then passes it as properties to the component responsible for rendering visual elements.

Over time, you might end up with a lot of container components in your app that all share similar state that needs to be passed to child components. This amounts to lots of code to pass property values around. For data that is truly global in your application, we can use context to access it.

Providing and consuming context

As your React application grows, it will use more components. Not only will it have more components, but the structure of your application will change so that the components are nested more deeply. The components that are nested at the deepest level still need to have data passed to them. Passing data from a parent component to a child component isn't a big deal. The challenge is when you have to start using components as indirection for passing data around your app.

For data that needs to make its way to any component in your app, you can create and use a context. There are two key concepts to remember when using contexts in React—providers and consumers. A context provider creates data and makes sure that it's available to any React components. A context consumer is a component that uses this data within the context.

You might be wondering whether or not context is just another way of saying **global data** in a React application. Essentially, this is exactly what contexts are used for. Using the React approach to wrap components with a context works better than creating global data because you have better control of how your data flows down through your components. For example, you can have nested contexts and a number of other advanced use cases. But, for now, let's just focus on simple usage.

Let's say that you have some application data that determines permissions for given application features. This data could be fetched from an API, or it could be hardcoded. In either case, the requirement is that you don't want to have to pass all of this permission data through the component tree. It would be nice if the permission data were just there for any component that needs it.

Starting at the very top of the component tree, let's look at `index.js`:

```
import * as React from "react";
import * as ReactDOM from "react-dom";
import { PermissionProvider } from "./PermissionContext";
import App from "./App";

const root =
  ReactDOM.createRoot(document.getElementById("root"));

root.render(
  <PermissionProvider>
    <App />
  </PermissionProvider>
);
```

The <App> component is the child of the <PermissionProvider> component. This means that the permission context has been provided to the <App> component and any of its children, all the way down the tree.

Let's take a look at the PermissionContext.js module where the permission context is defined:

```
import * as React from "react";

const { Provider, Consumer } =
  React.createContext("permissions");

class PermissionProvider extends React.Component {
  state = {
    first: true,
    second: false,
    third: true,
  };

  render() {
    return <Provider
      value={this.state}>{this.props.children}</Provider>;
  }
}

const PermissionConsumer = ({ name, children }) => (
  <Consumer>{(value) => value[name] && children}</Consumer>
);

export { PermissionProvider, PermissionConsumer };
```

The createContext() function is used to create the actual context. The return value is an object containing two components—Provider and Consumer. Next, there's a simple abstraction for the permission provider that's to be used throughout the app. The state contains the actual data that components might want to use. In this example, if the value is true, the feature should be displayed as normal. If it's false, then the feature doesn't have permission to render.

Here, the state is only set once; however, since our component is a regular React component, you could set the state in the same way you would set the state on any other component. The value that's rendered is the `<Provider>` component. This provides any children with context data, set via the `value` property.

Next, there's a small abstraction for permission consumers. Instead of having every component that needs to test for permissions implement the same logic over and over, the `PermissionConsumer` component can do it. The child of the component is always a function that takes the context data as an argument. In this example, the `PermissionConsumer` component has a `name` property for the name of the feature. This is compared with the value from the context and, if it's `false`, nothing is rendered.

Now let's look at the `App` component:

```
import * as React from "react";
import First from "./First";
import Second from "./Second";
import Third from "./Third";

export default () => (
  <>
    <First />
    <Second />
    <Third />
  </>
);
```

This component renders three components that are features and each needs to check for permissions. Without the context functionality of React, you would have to pass this data as a series of properties to each of these components through this component. If `<First>` had children or grandchildren that needed to check permissions, the same property-passing mechanism can get quite messy.

Now let's take a look at the <First> component (the <Second> and <Third> components are almost exactly the same):

```
import * as React from "react";
import { PermissionConsumer } from "./PermissionContext";

export default () => (
  <PermissionConsumer name="first">
    <div>
      <button>First</button>
    </div>
  </PermissionConsumer>
);
```

This is where the PermissionConsumer component is put to use. You just need to supply it with a name property, and the child component is the component that is rendered if the permission check passes. The <PermissionConsumer> component can be used anywhere, and there's no need to pass data in order to use it.

Here's what the rendered output of these three components looks like:

Figure 3.10 – The second button is hidden because it doesn't have the necessary permissions

The second component isn't rendered because its permission in the PermissionProvider component is set to false. Context should be used sparingly because it can lead to confusion about where data comes from and which components throughout your application rely on it. Often, you'll start out using state to manage data and then, later on, discover that you're passing this state to every component in your app.

To avoid this, you can refactor data that's shared by every component from state into context. Remember, context should be used sparingly. If you rely on context for accessing data too much, it's a good indication that your app has too much global data and should be revised. For the data that must be global, context is a good way to avoid too much property-passing code.

Summary

In this chapter, you learned about state and properties in React components. We started off by defining and comparing the two concepts. Then, we implemented several React components and manipulated their state, allowing you to dynamically update what the user sees on the screen. Next, you learned about properties by implementing code that passed property values from JSX to the component, in cases where the component only needs to display values instead of changing them.

Next, you were introduced to the concept of a container component, which is used to decouple data fetching from rendering content, leading to a clear separation of concerns. Finally, you learned about the new context API in React 16 and how to use it to avoid too many repetitive properties when you have global application data.

In the following chapter, you'll learn about the new React Hooks API and how it supports using functional components for everything, including state and lifecycle management.

Further reading

Visit the following links for more information:

- *Instance Properties*: `https://reactjs.org/docs/react-component.html#instance-properties-1`
- *Setting the Initial State*: `https://reactjs.org/docs/react-without-es6.html#setting-the-initial-state`
- *Context*: `https://reactjs.org/docs/context.html`
- *Spread syntax*: `https://developer.mozilla.org/en-US/docs/Web/JavaScript/Reference/Operators/Spread_syntax`

4
Getting Started with Hooks

One of the most anticipated new features of React is Hooks, an API that allows your functional components to "hook" into React functionality. The overarching motivation for this feature is to simplify your components. For example, forcing React developers to use classes to define their components leads to the overuse of wrapper components to pass state around their apps. With Hooks, you can stick with simple functions to implement your components and have a clear picture of how everything fits together.

In this chapter, we'll cover the following topics:

- Maintaining state using Hooks
- Performing initialization and cleanup actions
- Sharing data using context Hooks
- Using reducer Hooks to scale state management

Technical requirements

The code present in this chapter can be found at `https://github.com/PacktPublishing/React-and-React-Native-4th-Edition/tree/main/Chapter04`.

Maintaining state using Hooks

The first React Hook API that we'll look at is called `useState()`, which enables your functional React components to be stateful. Before Hooks were introduced to React, our only option for creating stateful components was to use a class so that we could access the `setState()` method. In this section, you'll learn how to initialize state values and change the state of a component using Hooks.

Initial state values

When our components are first rendered, they probably expect some state values to be set. This is called the initial state of the component, and we can use the `useState()` Hook to set the initial state. Let's take a look at an example:

```
import * as React from "react";

export default function App() {
  const [name] = React.useState("Adam");
  const [age] = React.useState(35);

  return (
    <>
      <p>My name is {name}</p>
      <p>My age is {age}</p>
    </>
  );
}
```

The `App` component is a functional React component, a function that returns JSX markup. But it's also now a stateful component, thanks to the `useState()` Hook. This example initializes two pieces of state, `name` and `age`. This is why there are two calls to `useState()`, one for each state value.

You can have as many pieces of state in your component as you need. The best practice is to have one call to `useState()` per state value. You could always define an object as the state of your component using only one call to `useState()`, but this complicates things because you have to access state values through an object instead of directly. Updating state values is also more complicated using this approach. When in doubt, use one `useState()` Hook per state value.

When we call `useState()`, we get an array returned to us. The first value of this array is the state value itself. Since we've used array-destructuring syntax here, we can call the value whatever we want; in this case, it is `name` and `age`. Both of these constants have values when the component is first rendered because we passed the initial state values for each of them to `useState()`. Here's what the page looks like when it's rendered:

My name is Adam

My age is 35

Figure 4.1 – Rendered output using values from state Hooks

Now that you've seen how to set the initial state values of your components, let's learn about updating these values.

Updating state values

React components use state for values that change over time. The state values used by components start off in one state, as we saw in the previous section, and then change in response to some event – for example, the server responds to an API request with new data, or the user has clicked a button or changed a form field.

With functional components that use the `useState()` Hook, state values are updated differently to class components that rely on the `setState()` method. Instead of using `setState()` to update every piece of component state, you have individual functions to set each state value. The `useState()` Hook returns an array. The first item is the state value and the second is the function used to update the value. Let's take a look at an example:

```
import * as React from "react";

function App() {
  const [name, setName] = React.useState("Adam");
  const [age, setAge] = React.useState(35);

  return (
    <>
      <section>
        <input
          value={name}
          onChange={(e) => setName(e.target.value)}
```

```
        />
        <p>My name is {name}</p>
      </section>
      <section>
        <input
          type="number"
          value={age}
          onChange={(e) => setAge(e.target.value)}
        />
        <p>My age is {age}</p>
      </section>
    </>
  );
}

export default App;
```

Just like the example from the *Initial state values* section, the `App` component in this example has two pieces of state – `name` and `age`. Unlike the previous example, this component uses two functions to update each piece of state. These are returned from the call to `useState()`. Let's take a closer look:

```
const [name, setName] = React.useState("Adam");
const [age, setAge] = React.useState(35);
```

Now, we have two functions – `setName()` and `setAge()` – that can be used to update the state of our component. Let's take a look at the text input field that updates the `name` state:

```
<section>
  <input
    value={name}
    onChange={(e) => setName(e.target.value)}
  />
  <p>My name is {name}</p>
</section>
```

Whenever the user changes the text in the <input> field, the `onChange` event is triggered. The handler for this event calls `setName()`, passing it `e.target.value` as an argument. The argument passed to `setName()` is the new state value of `name`. The succeeding paragraph shows that the text input is also updated with the new `name` value every time the user changes the text input.

Next, let's look at the `age` number input field and how this value is passed to `setAge()`:

```
<section>
  <input
    type="number"
    value={age}
    onChange={(e) => setAge(e.target.value)}
  />
  <p>My age is {age}</p>
</section>
```

The `age` field follows the exact same pattern as the `name` field. The only difference is that we've made the input a number type. Any time the number changes, `setAge()` is called with the updated value in response to the `onChange` event. The following paragraph shows that the number input is also updated with every change that is made to the `age` state.

Here is what the two inputs and their two corresponding paragraphs look like when they're rendered on the screen:

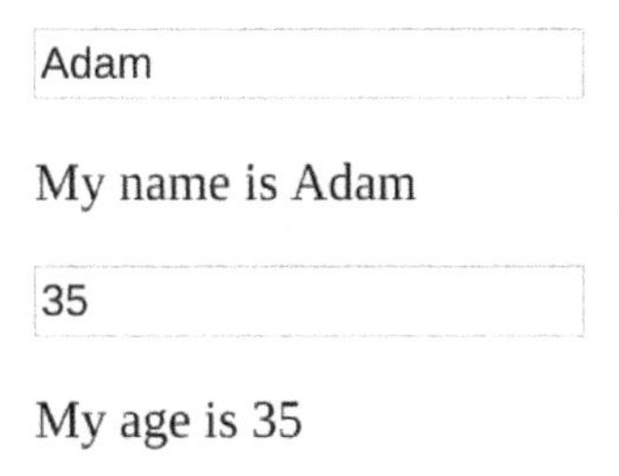

Figure 4.2 – Using Hooks to change state values

In this section, you learned about the `useState()` Hook, which is used to add state to functional React components. Each piece of state uses its own Hook and has its own value variable and its own setter function. This greatly simplifies accessing and updating state in your components. Any given state value should have an initial value so that the component can render correctly the first time. To re-render functional components that use state Hooks, you can use the setter functions that `useState()` returns to update your state values as needed.

The next Hook that you'll learn about is used to perform initialization and cleanup actions.

Performing initialization and cleanup actions

Often, our React components need to perform actions when the component is created. For example, a common initialization action is to fetch API data that the component needs. Another common action is to make sure that any pending API requests are canceled when the component is removed. In this section, you'll learn about the `useEffect()` Hook and how it can help you with these two scenarios. You'll also learn how to make sure that the initialization code doesn't run too often.

Fetching component data

The `useEffect()` Hook is used to run "side-effects" in your component. Another way to think about side-effect code is that functional components have only one job – returning JSX content to render. If the component needs to do something else, such as fetching API data, this should be done in a `useEffect()` Hook. For example, if you were to just make the API call as part of your component function, you would likely introduce race conditions and other difficult-to-fix buggy behavior.

Let's take a look at an example that fetches API data using Hooks:

```
import * as React from "react";

function fetchUser() {
  return new Promise((resolve) => {
    setTimeout(() => {
      resolve({ id: 1, name: "Adam" });
    }, 1000);
  });
}

function App() {
  const [id, setId] = React.useState("loading...");
  const [name, setName] = React.useState("loading...");

  React.useEffect(() => {
    fetchUser().then((user) => {
```

```
            setId(user.id);
            setName(user.name);
        });
    });

    return (
        <>
            <p>ID: {id}</p>
            <p>Name: {name}</p>
        </>
    );
}

export default App;
```

The `useEffect()` Hook expects a function as an argument. This function is called after the component finishes rendering, in a safe way that doesn't interfere with anything else that React is doing with the component under the covers. Let's look at the pieces of this example more closely, starting with the mock API function:

```
function fetchUser() {
    return new Promise((resolve) => {
        setTimeout(() => {
            resolve({ id: 1, name: "Adam" });
        }, 1000);
    });
}
```

The `fetchUser()` function returns a promise. The promise resolves a simple object with two properties, `id` and `name`. The `setTimeout()` function delays the promise resolution for 1 second, so this function is asynchronous, just as a normal `fetch()` call would be.

Next, let's look at the Hooks used by the `App` component:

```
const [id, setId] = React.useState("loading...");
const [name, setName] = React.useState("loading...");

React.useEffect(() => {
```

```
    fetchUser().then((user) => {
      setId(user.id);
      setName(user.name);
    });
  });
```

As you can see, we're using two Hooks in this component – `useState()` and `useEffect()`. Combining Hook functionality like this is powerful and encouraged. First, we set up the `id` and `name` states of the component. Then, `useEffect()` is used to set up a function that calls `fetchUser()` and sets the state of our component when the promise resolves.

Here is what the `App` component looks like when it's first rendered, using the initial state of `id` and `name`:

ID: loading...

Name: loading...

Figure 4.3 – Displaying the loading text until the data arrives

After 1 second, the promise returned from `fetchUser()` is resolved with data from the API, which is then used to update the ID and name states. This results in `App` being rerendered:

ID: 1

Name: Adam

Figure 4.4 – The state changes, removing the loading text and displaying returned values

There is a good chance that your users will navigate around your application while an API request is still pending. The `useEffect()` Hook can be used to deal with canceling these requests.

Canceling requests and resetting state

There's a good chance that, at some point, your users will navigate your app and cause components to unmount before responses to their API requests arrive. When this happens, an error occurs because the component will attempt to update the state values of a component that has been removed.

Thankfully, the `useEffect()` Hook has a mechanism to clean up things such as pending API requests when the component is removed. Let's take a look at an example of this in action:

```
import * as React from "react";
import { Promise } from "bluebird";

Promise.config({ cancellation: true });

function fetchUser() {
  return new Promise((resolve) => {
    setTimeout(() => {
      resolve({ id: 1, name: "Adam" });
    }, 1000);
  });
}

function User() {
  const [id, setId] = React.useState("loading...");
  const [name, setName] = React.useState("loading...");

  React.useEffect(() => {
    const promise = fetchUser().then((user) => {
      setId(user.id);
      setName(user.name);
    });

    return () => {
      promise.cancel();
    };
  });

  return (
    <>
      <p>ID: {id}</p>
      <p>Name: {name}</p>
    </>
```

```
  );
}

export default User;
```

This looks a lot like the component from the fetching component data example. It has the same state, it fetches data inside `useEffect()`, and it renders the same output. There are a couple of important differences though. Let's start by taking a closer look at the `useEffect()` Hook:

```
useEffect(() => {
  const promise = fetchUser().then((user) => {
    setId(user.id);
    setName(user.name);
  });

  return () => {
    promise.cancel();
  };
});
```

Just like in the fetching component data example, this effect creates a promise by calling the `fetchUser()` API function. It also returns a function, which React runs when the component is removed. In this example, the promise that is created by calling `fetchUser()` is canceled by calling `promise.cancel()`. This prevents the component from trying to update its state after it has been removed.

Another important difference compared with the preceding example is that here, we're using the Bluebird library for promises since they support cancellation. There are many other ways that you can "cancel" asynchronous operations in the function returned by the `useEffect()` Hook, but I found Bluebird to be well worth the added dependency for this added capability.

Now, let's look at the `App` component, which renders and removes the `User` component:

```
import * as React from "react";
import User from "./User";

const ShowHideUser = ({ show }) => (show ? <User /> :
  null);
```

```
function App() {
  const [show, setShow] = React.useState(false);

  return (
    <>
      <button onClick={() => setShow(!show)}>
        {show ? "Hide User" : "Show User"}
      </button>
      <ShowHideUser show={show} />
    </>
  );
}

export default App;
```

The `App` component renders a button that is used to toggle the `show` state. This state value determines whether or not the `User` component is rendered but by using the `ShowHideUser` convenience component. If `show` is `true`, `<User>` is rendered; otherwise, `User` is removed, triggering our `useEffect()` cleanup behavior.

Here's what the screen looks like when it first loads:

Show User

Figure 4.5 – A button used to initiate the state change

The `User` component isn't rendered because the `show` state of the `App` component is `false`. Try clicking on the show user button. This will change the `show` state and render the `User` component:

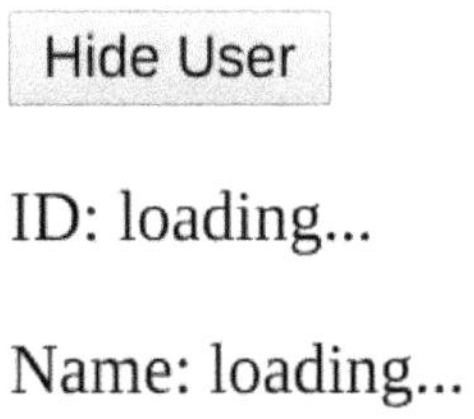

Hide User

ID: loading...

Name: loading...

Figure 4.6 – Displays the loading text when first clicked

The **loading...** strings are the two initial state values for the `id` and `name` states. These will be updated when the API promise resolves after 1 second:

Hide User

ID: 1

Name: Adam

Figure 4.7 – The loading strings are eventually replaced with new state data

You can click on the **Hide User** button once more to remove the `User` component. Now, click on the **Show User** button, and then click on **Hide User** before it finishes loading. Without the cleanup code that we added to `useEffect()`, this will trigger an error. In fact, you can test this by commenting out the call to `promise.cancel()`.

Effects are run by React after every render. This might not be what you want, especially if your effect is something that is relatively slow, such as an asynchronous network request. Instead, we want to call the API after the first render, and that's it. We'll take a look at how to do this next.

Optimizing side-effect actions

By default, React assumes that every effect that is run needs to be cleaned up. This typically isn't the case. For example, you might have specific property or state values that require cleanup when they change. You can pass an array of values to watch as the second argument to `useEffect()` – for example, if you have a resolved state that requires cleanup when it changes, you would write your effect code like this:

```
const [resolved, setResolved] = useState(false);
useEffect(() => {
  // ...the effect code...
  return () => {
    // ...the cleanup code that depends on "resolved"
  }
}, [resolved]);
```

In this code, the cleanup function will only ever run if the `resolved` state value changes. If the effect runs and the `resolved` state hasn't changed, then the cleanup code will not run. Another common case is to never run the cleanup code, except for when the component is removed. In fact, this is what we want to happen in the example from the previous section. Right now, the cleanup code runs after every render. This means that we're repeatedly fetching the user API data when all we really want is to fetch it once when the component is first mounted.

Let's make some modifications to the `User` component from the canceling requests example:

```
import * as React from "react";
import { Promise } from "bluebird";

Promise.config({ cancellation: true });

function fetchUser() {
  console.count("fetching user");
  return new Promise((resolve) => {
    setTimeout(() => {
      resolve({ id: 1, name: "Adam" });
    }, 1000);
  });
}

function User() {
  const [id, setId] = React.useState("loading...");
  const [name, setName] = React.useState("loading...");

  React.useEffect(() => {
    const promise = fetchUser().then((user) => {
      setId(user.id);
      setName(user.name);
    });

    return () => {
      promise.cancel();
    };
```

```
  }, []);

  return (
    <>
      <p>ID: {id}</p>
      <p>Name: {name}</p>
    </>
  );
}

export default User;
```

We've added a second argument to `useEffect()`, an empty array. This tells React that there are no values to watch and that we only want to run the cleanup code when the component is removed. We've also added `console.count('fetching user')` to the `fetchUser()` function. This makes it easier to look at the browser dev tools console and make sure that our component data is only fetched once. If you remove the `[]` argument that is passed to `useEffect()`, you'll notice that `fetchUser()` is called several times.

In this section, you learned about side effects in React components. Effects are an important concept, as they are the bridge between your React components and the outside world. One of the most common use cases for effects is to fetch data that the component needs, when it is first created, and then clean up after the component when it is removed.

Now, we're going to look at another way to share data with React components - context.

Sharing data using context Hooks

React applications often have a few pieces of data that are global in nature. This means that several components, possibly every component in an app, share this data - for example, information about the currently logged-in user might be used in several places. In cases like this, it makes sense to provide a context where this data can be easily accessed by components that are rendered in this context.

In this section, you'll learn how to consume context data using Hooks.

Sharing fetched data

Most of our components will directly fetch data that they and their children need. In other cases, our app has some API endpoint with data that is used by several components throughout the application. To share global data like this, you can use the React context API. As the name suggests, components that are rendered within a context are able to access the data provided by the context.

Let's build an example to help clarify what this means and how it relates to Hooks. Here is the `UserContext` context and the `UserProvider` component:

```
import * as React from "react";

export const UserContext = React.createContext();

function fetchUser() {
  return new Promise((resolve) => {
    setTimeout(() => {
      resolve({ id: 1, name: "Adam" });
    }, 1000);
  });
}

export function UserProvider({ children }) {
  const [user, setUser] = React.useState({ name: "..." });

  React.useEffect(() => {
    fetchUser().then((user) => {
      setUser(user);
    });
  }, []);

  return (
    <UserContext.Provider value={user}>
      {children}
    </UserContext.Provider>
  );
}
```

First, we have the `UserContext` object, created by calling the `createContext()` React API. Next, we have the mock API function, `fetchUser()`. Finally, we have the `UserProvider` component. The job of this component is to call the `fetchUser()` API and set the `user` state as the response from the API when it arrives. To do this, we're using the `useState()` and `useEffect()` Hooks.

This component renders the `<UserContext.Provider>` component, passing in any children it receives. The `value` property is then made available to any child components of `UserProvider`. In this case, the value is the state that is set by calling the `fetchUser()` API. We've set ourselves up to be able to pass the user value to any component of our application. Let's see how this is done by creating a simple `App` component with three pages on it:

```
import * as React from "react";
import { UserProvider } from "./UserContext";
import { Page1, Page2, Page3 } from "./Pages";

function ChoosePage({ page }) {
  const Page = [Page1, Page2, Page3][page];
  return <Page />;
}

function App() {
  const [page, setPage] = React.useState(0);

  return (
    <UserProvider>
      <button onClick={() => setPage(0)} disabled=
        {page === 0}>
        Page 1
      </button>
      <button onClick={() => setPage(1)} disabled=
        {page === 1}>
        Page 2
      </button>
```

```
        <button onClick={() => setPage(2)} disabled=
          {page === 2}>
          Page 3
        </button>
        <ChoosePage page={page} />
      </UserProvider>
    );
  }

export default App;
```

The App component renders three buttons that, when clicked, render their corresponding page component. The page state is used to control the page that is displayed and defaults to 0. When App is first rendered, the Page1 component is rendered. This happens with the help of ChoosePage, which renders the correct page based on the page state that is passed to it. Here's what you'll see when the page state first loads:

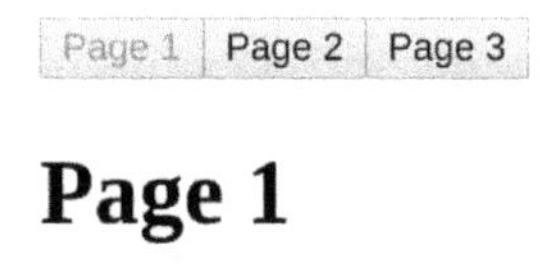

Figure 4.8 – The default view of our page

The **Page 1** button is disabled because it is the currently active page. There's an ellipsis following the **Logged in as** message at the bottom of the page. This is because the UserProvider component is waiting for the fetchUser() API call to respond. When the response arrives and the context data is updated, the Page1 component is updated:

Figure 4.9 – The state changes when the fetch call resolves, updating the user name

Last but not least, let's take a look at the page components that use context Hooks:

```
import * as React from "react";
import { UserContext } from "./UserContext";

function Username() {
  const user = React.useContext(UserContext);
  return (
    <p>
      Logged in as <strong>{user.name}</strong>
    </p>
  );
}

export function Page1() {
  return (
    <>
      <h1>Page 1</h1>
      <Username />
    </>
  );
}

export function Page2() {
  return (
    <>
      <h1>Page 2</h1>
      <Username />
    </>
  );
}

export function Page3() {
  return (
    <>
      <h1>Page 3</h1>
      <Username />
```

```
    </>
  );
}
```

All three page components look pretty much the same, except for the text used in each. Let's focus on the `Username` component that is used by each page:

```
function Username() {
  const user = React.useContext(UserContext);
  return (
    <p>
      Logged in as <strong>{user.name}</strong>
    </p>
  );
}
```

This is where the `useContext()` Hook is used. The `user` context value is actually the state that is set by the `UserProvider` component when the API call responds. This means that the `user` context value is updated by the `useContext()` Hook whenever the `user` value changes.

Another important idea from this example is that the page components (`Page1`, `Page2`, and `Page3`) have no knowledge of this global user data. Instead of having to pass data down from the top-level component as property values, we can rely on `useContext()` when we need access to global data, no matter how deeply nested the component is in our JSX markup. For components that have nothing to do with the data, such as the page components in this example, there's no need to touch them.

Updating stateful context data

Global data that is shared throughout your application isn't limited to read-only API response data. Sometimes, components themselves need to update global state values. To enable this capability, we need to pass not only data from context producers but also a mechanism to update the data. Since data stored in a context provider is a state created with `useState()`, we can just pass along the setter function, along with the state value.

Let's illustrate these ideas by extending the sharing fetched data example. Instead of a user context, we'll add a `status` context. This way, components that are rendered within this context will have access to the status state value and the `status` state setter function. Here's what the `StatusProvider` component looks like:

```
import * as React from "react";

export const StatusContext = React.createContext();

export function StatusProvider({ children }) {
  const value = React.useState("set a status");

  return (
    <StatusContext.Provider value={value}>
      {children}
    </StatusContext.Provider>
  );
}
```

The `StatusProvider` component has a status state with a default string value. Recall that `useState()` returns an array of state value and a state setter function. This array is then passed to the `value` property of `<StatusContext.Provider>`. Now, let's take a look at the page components that display and update the status context data:

```
import * as React from "react";
import { StatusContext } from "./StatusContext";

function SetStatus() {
  const [status, setStatus] =
    React.useContext(StatusContext);
  return (
    <input
      value={status}
      onChange={(e) => setStatus(e.target.value)}
    />
  );
}
```

```
export function Status() {
  const [status] = React.useContext(StatusContext);
  return <p>{status}</p>;
}

export function Page1() {
  return (
    <>
      <h1>Page 1</h1>
      <SetStatus />
    </>
  );
}

export function Page2() {
  return (
    <>
      <h1>Page 2</h1>
    </>
  );
}

export function Page3() {
  return (
    <>
      <h1>Page 3</h1>
      <SetStatus />
    </>
  );
}
```

Let's take a closer look at the two utility components that consume context data with `useContext()`:

```
function SetStatus() {
  const [status, setStatus] =
    React.useContext(StatusContext);
  return (
    <input
      value={status}
      onChange={(e) => setStatus(e.target.value)}
    />
  );
}

export function Status() {
  const [status] = React.useContext(StatusContext);
  return <p>{status}</p>;
}
```

The `SetStatus` component is used to render an input so that the user can provide new values for the `status` context. When they do, the `setStatus()` function that comes from the context data array is used to update the context state. The `Status` component only renders status, so it doesn't need the `setStatus()` function that comes from `useContext()`. The `Page2` component doesn't render the `SetStatus` component, but `Page1` and `Page2` do.

The `Status` component is used by the `App` component to display status on every page, including `Page2`. Let's see these pages in action now. Here is what the first page looks like when it first loads, using the default status context:

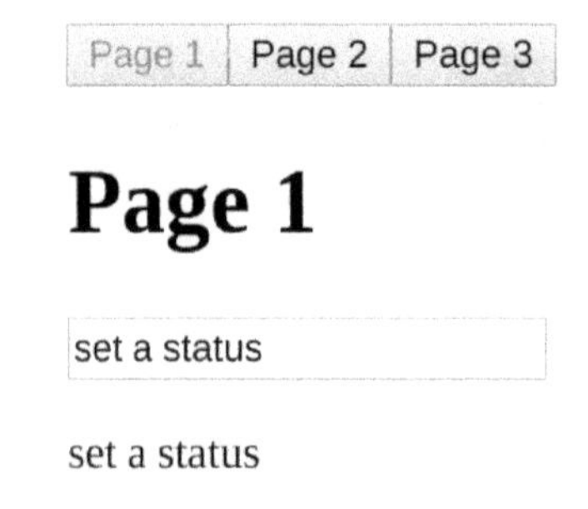

Figure 4.10 – Showing what the user is typing

The text input that sets the status is part of the `Page1` component. The succeeding status label shows that the text input that displays the status is part of the `App` component and will be rendered on every page. Let's try changing the status:

Figure 4.11 – The display changes as the user types

The `setStatus()` function that was passed in context data is used to update the status state in the `StatusProvider` component. The new context data is propagated throughout the application components that use it, any time it changes. Let's see what the second page looks like after we've updated the status:

Figure 4.12 – The status update is visible on page 2

The `Page2` component doesn't use the `SetStatus` component, which is why there's no input shown here. But the status label that is rendered by the `App` component hasn't changed. Lastly, let's take a look at the third page:

Page 1 Page 2 Page 3

Page 3

status updated

status updated

Figure 4.13 – Status changes are also reflected on page 3

As expected, the updated status context data is reflected here as well. In fact, since `Page3` uses the `SetStatus` component, you can update the status again and navigate the pages again. The result will be the same, since the same mechanics are in place.

This section showed you how to create a context for global data that various components in your application need to share. One common scenario is an API endpoint with data that most components in the application need access to. You can implement a context provider component that performs this API data fetch and then shares it with other components. The components that require this global data can use the `useContext()` Hook, which feels a lot like using the `useState()` Hook.

You also learned that context data can be changed by different components. This involves passing a state setting function as part of the context data so that components can use it to update the context value. In the next section, we'll look at using reducer Hooks to help simplify complex state management.

Using reducer Hooks to scale state management

The `useState()` Hook is a great way to manage the state of your component. It can become a challenge to use this Hook when your component has a lot of related pieces of state. You end up with a lot of setter functions that you need to call individually, once you've figured out how a change in one state value affects another state value. With reducers, you have one `dispatch()` function that's used to update the state of your component.

In this section, you'll learn about the basics of reducer actions and how they update the state of your component. Then, we'll look at a more in-depth example that shows you how to handle updating state values that depend on other state values.

Using reducer actions

A reducer function in a React application is a function that takes the current state, an `action`, and any other arguments that are needed to update the state. It returns the new state of the component. The `action` argument tells the reducer function what new state to return and is often used in a `switch` statement. Let's look at an example now:

```
function reducer(state, action) {
  switch (action.type) {
    case "changeName":
      return { ...state, name: action.value };
```

```
    case "changeAge":
      return { ...state, age: action.value };
    default:
      throw new Error('${action.type} is not a valid
        action');
  }
}

function App() {
  const [{ name, age }, dispatch] =
    React.useReducer(reducer, {
    name: "",
    age: "",
  });

  return (
    <>
      <input
        placeholder="Name"
        value={name}
        onChange={(e) =>
          dispatch({ type: "changeName", value:
            e.target.value })
        }
      />
      <p>Name: {name}</p>
      <input
        placeholder="Age"
        type="number"
        value={age}
        onChange={(e) =>
          dispatch({ type: "changeAge", value:
            e.target.value })
        }
      />
```

```
      <p>Age: {age}</p>
    </>
  );
}
```

Here, we have an `App` component that renders two fields and two labels. When the text value changes, it should update the corresponding label value. This is done by using two pieces of state, one for each field. Let's take a closer look at how state is set up with the `useReducer()` Hook:

```
const [{ name, age }, dispatch] =
  React.useReducer(reducer, {
  name: "",
  age: "",
});
```

The `useReducer()` function takes two arguments – the reducer function that updates the state, and the initial state of the component. The return value of `useReducer()` is an array, with the state as the first element and the dispatcher function as the second. When we use reducers, we only have one object as the state of the component, instead of several smaller, unrelated state values. This is why we're destructuring the `state` object into `name` and `age` constants. Now, let's take a look at the reducer function itself:

```
function reducer(state, action) {
  switch (action.type) {
    case "changeName":
      return { ...state, name: action.value };
    case "changeAge":
      return { ...state, age: action.value };
    default:
      throw new Error('${action.type} is not a valid
        action');
  }
}
```

The `state` argument is the current state of the component. The `action` argument is the argument that's passed to `dispatch()`. The `action.type` value is used to determine what to do. This reducer only has two possible actions – `changeName` and `changeAge`. Based on this, we use the object spread operator to return a new state object, made from the existing state and the updated state object values. In this case, based on the `action.type` value, either the `name` or `age` state values will be updated.

It's also important to have a default handler in place that throws an error when an unexpected action is passed to the reducer. It's highly likely that you will get this wrong at some point, and it's better to have the reducer complain loudly about the invalid action than to have to figure out why your component has the wrong state set on it.

Here is what the screen looks like when `App` is first rendered:

Name

Name:

Age

Age:

Figure 4.14 – The default input values

Here's what you'll see when you enter some text into these two inputs:

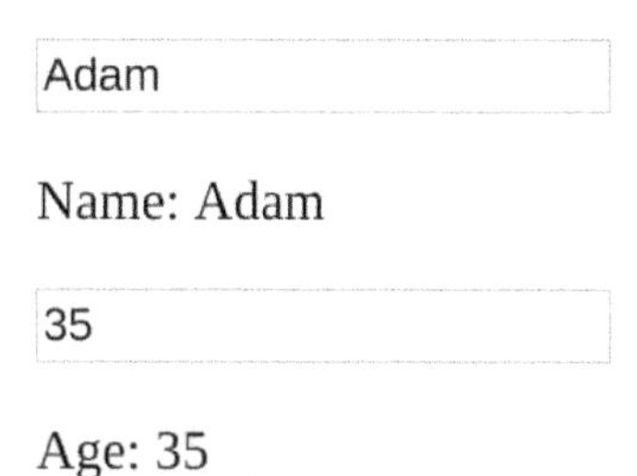

Figure 4.15 – The updated labels once text is added

This example used a reducer function to update two unrelated pieces of state. In other words, you probably could have used the `useState()` Hook just as easily. However, now that you have an idea of what reducers are and how they handle different actions that are dispatched to them, you're ready to look at a more complex example that involves state values that depend on other state values.

Handling state dependencies

When our components have one piece of state that depends on another, it's difficult to use the `useState()` Hook. This Hook comes with the assumption that when a state needs to be updated, it's one piece at a time. In real applications, there are often scenarios where updating one piece of state means that another piece of state needs to be updated as well, based on this new value.

Let's look at an example that allows the user to select an item and the quantity of that item. It then shows the cost. This means that whenever the quantity or item fields change, the total must also change. Here's the reducer code:

```
const initialState = {
  options: [
    { id: 1, name: "First", value: 10 },
    { id: 2, name: "Second", value: 50 },
    { id: 3, name: "Third", value: 200 },
  ],
  quantity: 1,
  selected: 1,
};

function reduceButtonStates(state) {
  return {
    ...state,
    decrementDisabled: state.quantity === 0,
    incrementDisabled: state.quantity === 10,
  };
}

function reduceTotal(state) {
  const option = state.options.find(
    (option) => option.id === state.selected
  );
  return { ...state, total: state.quantity * option.value
    };
}

function reducer(state, action) {
```

```
    let newState;
    switch (action.type) {
     case "init":
       newState = reduceTotal(state);
       return reduceButtonStates(newState);
     case "decrementQuantity":
       newState = { ...state, quantity: state.quantity - 1 };
       newState = reduceTotal(newState);
       return reduceButtonStates(newState);
     case "incrementQuantity":
       newState = { ...state, quantity: state.quantity + 1 };
       newState = reduceTotal(newState);
       return reduceButtonStates(newState);
     case "selectItem":
       newState = { ...state, selected: Number(action.id) };
       return reduceTotal(newState);
     default:
       throw new Error('${action.type} is not a valid
         action');
    }
  }
```

Here's the App component that uses the reducer:

```
  export default function App() {
    const [
      {
        options,
        selected,
        quantity,
        total,
        decrementDisabled,
        incrementDisabled,
      },
      dispatch,
    ] = React.useReducer(reducer, initialState);
```

```
React.useEffect(() => {
  dispatch({ type: "init" });
}, []);

return (
  <>
    <section>
      <button
        disabled={decrementDisabled}
        onClick={() => dispatch({ type:
          "decrementQuantity" })}
      >
        -
      </button>
      <button
        disabled={incrementDisabled}
        onClick={() => dispatch({ type:
          "incrementQuantity" })}
      >
        +
      </button>
      <input readOnly value={quantity} />
    </section>
    <section>
      <select
        value={selected}
        onChange={(e) =>
          dispatch({ type: "selectItem", id:
            e.target.value })
        }
      >
        {options.map((o) => (
          <option key={o.id} value={o.id}>
            {o.name}
          </option>
        ))}
```

```
        </select>
      </section>
      <section>
        <strong>{total}</strong>
      </section>
    </>
  );
}
```

Before jumping into code explanations, let's see what this code actually does. Here's what you'll see when the screen first loads:

Figure 4.16 – The default field values when the screen first loads

By default, the quantity is set to 1 and the **First** item is selected. The total cost is displayed beneath the two fields. When the page first loads, the total is **10**, since the cost of the **First** item is **10** and the quantity is set to **1**. Let's try changing the quantity value, using the increment and decrement buttons beside it:

Figure 4.17 – Incrementing the numeric field value

Here, we've changed the quantity to **5**. As you can see, the total reflects this quantity by changing to **50**. The quantity state has minimum (0) and maximum (10) restrictions, so if you increase the quantity value to **10**, the increment button is disabled:

Figure 4.18 – The maximum allowed value

If you change the selected item, the total is reflected, based on the current quantity value:

Figure 4.19 – Changing the multiplier value

This example has several pieces of state that depend on one another in moderately complex ways. This is a perfect opportunity to put the `useReducer()` Hook into action. Let's break down what's going on in the code. We'll start by looking at the initial state:

```
const initialState = {
  options: [
    { id: 1, name: "First", value: 10 },
    { id: 2, name: "Second", value: 50 },
    { id: 3, name: "Third", value: 200 },
  ],
  quantity: 1,
  selected: 1,
};
```

The `options` array is the items that the user can select from; the initial `quantity` value is `1`, and the `selected` value represents which item is selected. Later on, this component will set several other state values, but these are all that are needed for the initial render. Next, let's take a closer look at the reducer functions that maintain the state of this component:

```
function reduceButtonStates(state) {
  return {
    ...state,
    decrementDisabled: state.quantity === 0,
    incrementDisabled: state.quantity === 10,
  };
}

function reduceTotal(state) {
  const option = state.options.find(
    (option) => option.id === state.selected
  );
```

```
    return { ...state, total: state.quantity * option.value
      };
  }

  function reducer(state, action) {
    let newState;
    switch (action.type) {
      case "init":
        newState = reduceTotal(state);
        return reduceButtonStates(newState);
      case "decrementQuantity":
        newState = { ...state, quantity: state.quantity - 1 };
        newState = reduceTotal(newState);
        return reduceButtonStates(newState);
      case "incrementQuantity":
        newState = { ...state, quantity: state.quantity + 1 };
        newState = reduceTotal(newState);
        return reduceButtonStates(newState);
      case "selectItem":
        newState = { ...state, selected: Number(action.id) };
        return reduceTotal(newState);
      default:
        throw new Error('${action.type} is not a valid
          action');
    }
  }
```

The `reducer()` function is passed to `useReducer()` and is responsible for handling different action paths. This particular reducer handles the following actions:

- `init`: When the component first mounts.
- `decrementQuantity`: The decrement quantity button was pressed.
- `incrementQuantity`: The increment quantity button was pressed.
- `selectItem`: The selected item was changed.

Every one of these actions has the potential to change the total state, which is why the code to compute the total was moved into its own function - `reduceTotal()`. For example, if the quantity changes or the item changes, we need to compute a new total. When the component first mounts, we also need to compute the total because we don't want to have a default state for something that's derived from other state values. Instead, we introduce the `init` action and use the `useEffect()` Hook to call it once when the component is first mounted.

The state of the increment and decrement buttons is dependent on the quantity value. So, the `incrementDisabled` and `decrementDisabled` state values are computed in the `reduceButtonStates()` function, which is used by the `init`, `decrementQuantity`, and `incrementQuantity` actions.

At first glance, it might seem like there's a lot going on in the `reducer()` function, and you'd be right - there is. But in this example, the goal is to keep related state operations close to one another, since they're related. The perfect place to do this is in a reducer function. Developers look at our code and follow the action flow without much trouble. We also managed to factor out common reducer behavior into its own functions. All of this results in a functional component that doesn't have to directly perform any complex state updates. Instead, it just needs to make `dispatch()` calls, keeping the component itself focused on markup and event handling.

In this section, you learned that the `useReducer()` Hook is similar to the `useState()` Hook in that they are both React state management APIs. Using a reducer function is helpful when you want to keep your component state together as a single object so that you can update it more easily when the updates are complex due to dependencies.

Summary

This chapter introduced you to the new React Hooks API. You started by using the `useState()` Hook, which is fundamental for using state in functional React components. Then, you learned about `useEffect()`, which enables life cycle management in functional React components, such as fetching API data when a component is mounted and cleaning up any pending async operations when it is removed. Then, you learned how to use the `useContext()` Hook in order to access global application data. Lastly, you learned about the `useReducer()` Hook - an effective replacement for `useState()` when your component state grows too big or too complex for `useState()`.

In the following chapter, you'll learn about event handling in React components.

5
Event Handling, the React Way

The focus of this chapter is **event handling**. React has a unique approach to handling events – declaring event handlers in JSX. We'll get things started by looking at how event handlers for particular elements are declared in JSX. Then, we'll learn about binding handler context and parameter values. Next, we'll implement inline and higher-order event handler functions.

Then, you'll learn how React actually maps event handlers to DOM elements under the hood. Finally, you'll learn about the synthetic events that React passes to event handler functions and how they're pooled for performance purposes. Once you've completed this chapter, you'll be comfortable implementing event handlers in your React components. At that point, your applications come to life for your users because they are then able to interact with them.

The following topics are covered in this chapter:

Declaring event handlers

Using event handler context and parameters

Declaring inline event handlers

Binding handlers to elements

Using synthetic event objects

Understanding event pooling

Technical requirements

The code present in this chapter can be found at `https://github.com/PacktPublishing/React-and-React-Native-4th-Edition/tree/main/Chapter05`.

Declaring event handlers

The differentiating factor with event handling in React components is that it's declarative. Contrast this with something such as jQuery, where you have to write imperative code that selects the relevant DOM elements and attaches event handler functions to them.

The advantage of the declarative approach to event handlers in JSX markup is that they're part of the UI structure. Not having to track down code that assigns event handlers is mentally liberating.

In this section, you'll write a basic event handler so that you can get a feel for the declarative event-handling syntax found in React applications. Then, you'll learn how to use generic event handler functions.

Declaring handler functions

Let's take a look at a basic component that declares an event handler for the click event of an element:

```
import * as React from "react";

class MyButton extends React.Component {
  onClick() {
    console.log("clicked");
  }

  render() {
    return (
      <button onClick={this.onClick}>{this.props.children}
        </button>
    );
  }
}

export default MyButton;
```

The event handler `this.onClick()` function is passed to the `onClick` property of the `<button>` element. By looking at this markup, you can see exactly which code will run when the button is clicked.

> **Note**
>
> View the official React documentation for the full list of supported event property names at `https://reactjs.org/docs/events.html#supported-events`.

Next, let's take a look at how to respond to more than one type of event using different event handlers with the same element.

Multiple event handlers

What I really like about the declarative event handler syntax is that it's easy to read when there's more than one handler assigned to an element. Sometimes, for example, there are two or three handlers for an element. Imperative code is difficult to work with for a single event handler, let alone several of them. When an element needs more handlers, it's just another JSX attribute. This scales well from a code-maintainability perspective, as this example shows:

```
import * as React from "react";

class MyInput extends React.Component {
  onChange() {
    console.log("changed");
  }

  onBlur() {
    console.log("blured");
  }

  render() {
    return <input onChange={this.onChange}
      onBlur={this.onBlur} />;
  }
}

export default MyInput;
```

This <input> element could have several more event handlers and the code would be just as readable.

As you keep adding more event handlers to your components, you'll notice that a lot of them do the same thing. Next, you'll learn how to share generic handler functions across components.

Importing generic handlers

Any React application is likely going to share the same event-handling logic for different components. For example, in response to a button click, the component should sort a list of items. It's these types of generic behaviors that belong in their own modules so that several components can share them.

Let's implement a component that uses a generic event handler function:

```
import * as React from "react";
import reverse from "./reverse";

class MyList extends React.Component {
  state = {
    items: ["Angular", "Ember", "React"],
  };

  onReverseClick = reverse.bind(this);

  render() {
    const {
      state: { items },
      onReverseClick,
    } = this;

    return (
      <section>
        <button onClick={onReverseClick}>Reverse</button>
        <ul>
          {items.map((v, i) => (
            <li key={i}>{v}</li>
          ))}
```

```
        </ul>
      </section>
    );
  }
}

export default MyList;
```

Let's walk through what's going on here, starting with the imports. You're importing a function called `reverse()`. This is the generic event handler function that you're using with your `<button>` element. When it's clicked, the list should reverse its order.

The `onReverseClick` method actually calls the generic `reverse()` function. It is created using `bind()` to bind the context of the generic function to this component instance.

Finally, looking at the JSX markup, you can see that the `onReverseClick()` function is used as the handler for the button click.

So, how does this work exactly? You have a generic function that somehow changes the state of this component because you bound context to it? Well, pretty much, yes – that's it. Let's look at the generic function implementation now:

```
export default function reverse() {
  this.setState(this.state.items.reverse());
}
```

This function depends on a `this.state` property and an `items` array within the state. The key is that the state is generic; an application can have many components with an `items` array in its state.

Here's what our rendered list looks like:

Reverse

- Angular
- Ember
- React

Figure 5.1 – The rendered list of values

As expected, clicking on the button causes the list to sort, using your generic `reverse()` event handler:

Reverse

- React
- Ember
- Angular

Figure 5.2 – The list of values reversed

In this section, you learned how to declare event handler functions for your JSX elements. You then learned how to assign more than one event handler to an element and import and use generic handler functions. Next, you'll learn how to bind the context and the argument values of event handler functions.

Using event handler context and parameters

In this section, you'll learn about React components that bind their event handler contexts and how you can pass data into event handlers. Having the right context is important for React event handler functions because they usually need access to component properties or state. Being able to parameterize event handlers is also important because they don't pull data out of DOM elements.

Getting component data

In this section, you'll learn about scenarios where the handler needs access to component properties, along with argument values. You'll render a custom list component that has a click event handler for each item in the list. The component is passed an array of values, as follows:

```
import * as React from "react";
import * as ReactDOM from "react-dom";
import MyList from "./MyList";

const items = [
  { id: 0, name: "First" },
  { id: 1, name: "Second" },
  { id: 2, name: "Third" },
];

const root =
```

```
  ReactDOM.createRoot(document.getElementById("root"));
root.render(<MyList items={items} />);
```

Each item in the list has an `id` property, which is used to identify the item. You'll need to be able to access this ID when the item is clicked on in the UI so that the event handler can work with the item.

Here's what the `MyList` component implementation looks like:

```
import * as React from "react";

class MyList extends React.Component {
  constructor() {
    super();
    this.onClick = this.onClick.bind(this);
  }

  onClick(id) {
    const { name } = this.props.items.find((i) => i.id ===
      id);
    console.log("clicked", '"${name}"');
  }

  render() {
    return (
      <ul>
        {this.props.items.map(({ id, name }) => (
          <li key={id} onClick={this.onClick.bind(null,
            id)}>
            {name}
          </li>
        ))}
      </ul>
    );
  }
}

export default MyList;
```

Here is what the rendered list looks like:

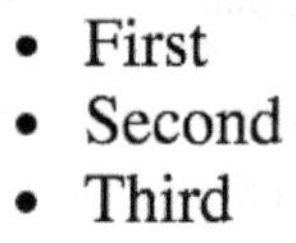

- First
- Second
- Third

Figure 5.3 – The rendered list of values

You have to bind the event handler context, which is done in the constructor. If you look at the `onClick()` event handler, you can see that it needs access to the component so that it can look up the clicked item in `this.props.items`. Also, the `onClick()` handler is expecting an `id` parameter. If you take a look at the JSX content of this component, you can see that calling `bind()` supplies the argument value for each item in the list. This means that when the handler is called in response to a click event, the `id` of the item is already provided.

This approach to parameterized event handling is quite different from prior approaches – for example, I used to rely on getting parameter data from the DOM element itself. This works well when you only need one event handler, and it can extract the data it needs from the event argument. This approach also doesn't require setting up several new functions by iterating over a collection and calling `bind()`.

And therein lies the trade-off. React applications avoid touching the DOM because it is really just a render target for React components. If you can write code that doesn't introduce explicit dependencies to DOM elements, your code will be portable. This is what you've accomplished with the event handler in this example.

> **Note**
>
> If you're concerned about the performance implications of creating a new function for every item in a collection, don't be. You're not going to render thousands of items on the page at a time. Benchmark your code, and if it turns out that `bind()` is the slowest part, then you probably have a really fast application.

In the next section, you'll learn how to build event handler functions on the fly using higher-order functions.

Higher-order event handlers

A **higher-order function** is a function that returns a new function. Sometimes, higher-order functions take functions as arguments too. In the *Getting component data* example, you used `bind()` to bind the context and argument values of your event handler functions. Higher-order functions that return event handler functions are another technique. The main advantage of this technique is that you don't have to call `bind()` several times. Instead, you just call the function where you want to bind parameters to the function.

Let's look at an example component:

```
import * as React from "react";

class App extends React.Component {
  state = {
    first: 0,
    second: 0,
    third: 0,
  };

  onClick = (name) => () => {
    this.setState((state) => ({
      ...state,
      [name]: state[name] + 1,
    }));
  };

  render() {
    const { first, second, third } = this.state;

    return (
      <>
        <button onClick={this.onClick("first")}>
          First {first}</button>
        <button onClick={this.onClick("second")}>
          Second {second}
        </button>
        <button onClick={this.onClick("third")}>
          Third {third}
```

```
            </button>
          </>
        );
    }
}

export default App;
```

This component renders three buttons and has three pieces of state – a counter for each button. The `onClick()` function is automatically bound to the component context because it's defined as an arrow function. It takes a `name` argument and returns a new function. The function that is returned uses this `name` value when called. It uses computed property syntax (variables inside `[]`) to increment the state value for the given name. Here's what that component content looks like after each button has been clicked a few times:

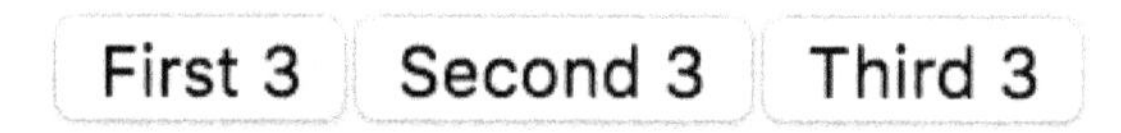

Figure 5.4 – Three buttons rendered

In this section, you learned how to make your event handler functions interact with your component data. If you have a class-based component, you can bind your function context to the component class so that you have direct access to the component state and properties. You also learned that higher-order functions are another option for generating distinct callback functions by passing an argument to the higher-order function.

In the next section, you'll learn about inline event handler functions.

Declaring inline event handlers

The typical approach to assigning handler functions to JSX properties is to use a named function. However, sometimes, you might want to use an inline function, where the function is defined as part of the markup. This is done by assigning an arrow function directly to the event property in the JSX markup:

```
import * as React from "react";

class MyButton extends React.Component {
  render() {
    return (
```

```
      <button onClick={(e) => console.log("clicked", e)}>
        {this.props.children}
      </button>
    );
  }
}

export default MyButton;
```

The main use of inlining event handlers like this is when you have a static parameter value that you want to pass to another function. In this example, you're calling `console.log()` with the clicked string. You could have set up a special function for this purpose outside of the JSX markup by creating a new function using `bind()`, or by using a higher-order function. But then you would have to think of yet another name for yet another function. Inlining is just easier sometimes.

Next, you'll learn about how React binds handler functions to the underlying DOM elements in the browser.

Binding handlers to elements

When you assign an event handler function to an element in JSX, React doesn't actually attach an event listener to the underlying DOM element. Instead, it adds the function to an internal mapping of functions. There's a single event listener on the document for the page. As events bubble up through the DOM tree to the document, the React handler checks to see whether any components have matching handlers. The process is illustrated here:

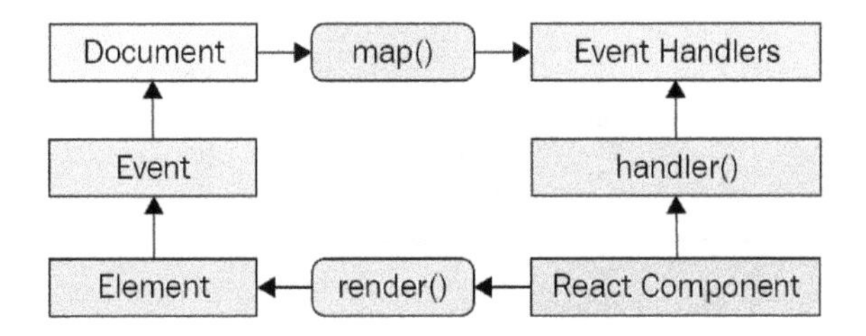

Figure 5.5 – The event handler cycle

Why does React go to all of this trouble, you might ask? It's the same principle that I've been covering in the last few chapters – keep the declarative UI structures separated from the DOM as much as possible.

For example, when a new component is rendered, its event handler functions are simply added to the internal mapping maintained by React. When an event is triggered and it hits the document object, React maps the event to the handlers. If a match is found, it calls the handler. Finally, when the React component is removed, the handler is simply removed from the list of handlers.

None of these DOM operations actually touch the DOM. It's all abstracted by a single event listener. This is good for performance and the overall architecture (keep the render target separate from the application code).

In the following section, you'll learn about the synthetic event implementation used by React to ensure good performance and safe asynchronous behavior.

Using synthetic event objects

When you attach an event handler function to a DOM element using the native `addEventListener()` function, the callback will get an event argument passed to it. Event handler functions in React are also passed an event argument, but it's not the standard Event instance. It's called `SyntheticEvent`, and it's a simple wrapper for native event instances.

Synthetic events serve two purposes in React:

- They provide a consistent event interface, normalizing browser inconsistencies.
- They contain information that's necessary for propagation to work.

Here's a diagram of the synthetic event in the context of a React component:

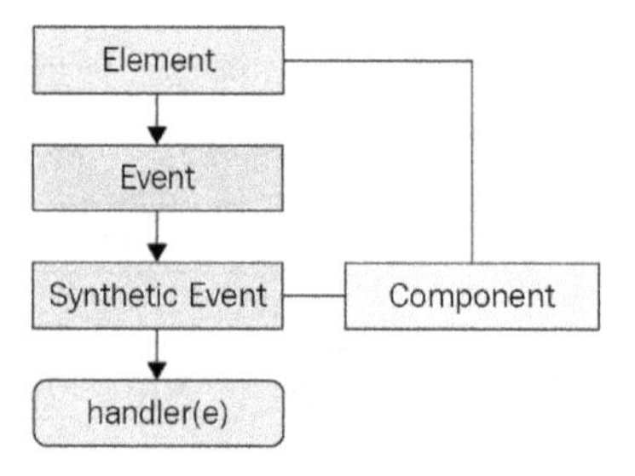

Figure 5.6 – How synthetic events are created and processed

When a DOM element that is part of a React component dispatches an event, React will handle the event because it sets up its own listeners for them. Then, it will either create a new synthetic event or reuse one from the pool, depending on availability. If there are any event handlers declared for the component that match the DOM event that was dispatched, they will run with the synthetic event passed to them.

In the next section, you'll see how these synthetic events are pooled for performance reasons and the implications of this on asynchronous code.

Understanding event pooling

One challenge of wrapping native event instances is that it can cause performance issues. Every synthetic event wrapper that's created will also need to be garbage-collected at some point, which can be expensive in terms of CPU time.

> **Note**
>
> When the garbage collector is running, none of your JavaScript code is able to run. This is why it's important to be memory-efficient; frequent garbage collection means less CPU time for code that responds to user interactions.

For example, if your application only handles a few events, this wouldn't matter much. But even by modest standards, applications respond to many events, even if the handlers don't actually do anything with them. This is problematic if React constantly has to allocate new synthetic event instances.

React deals with this problem by allocating a synthetic instance pool. Whenever an event is triggered, it takes an instance from the pool and populates its properties. When the event handler has finished running, the synthetic event instance is released back into the pool, as shown here:

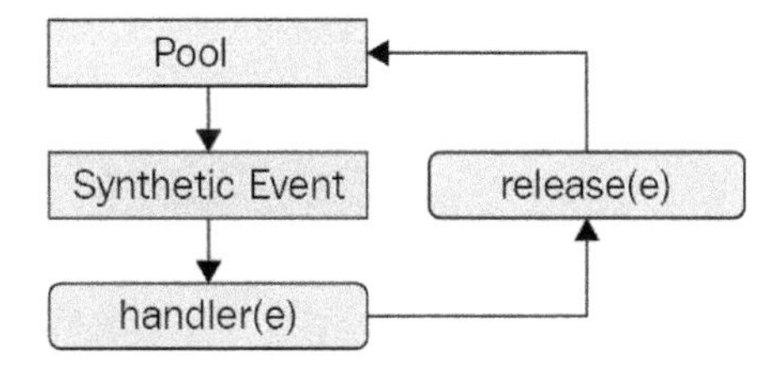

Figure 5.7 – Synthetic events are reused to save memory resources

This prevents the garbage collector from running frequently when a lot of events are triggered. The pool keeps a reference to the synthetic event instances, so they're never eligible for garbage collection. React never has to allocate new instances either.

However, there is one *gotcha* that you need to be aware of. It involves accessing the synthetic event instances from asynchronous code in your event handlers. This is an issue because, as soon as the handler has finished running, the instance goes back into the pool. When it goes back into the pool, all of its properties are cleared.

Here's an example that shows how this can go wrong:

```
import * as React from "react";

function fetchData() {
  return new Promise((resolve) => {
    setTimeout(() => {
      resolve();
    }, 1000);
  });
}

class MyButton extends React.Component {
  onClick(e) {
    const style = e.currentTarget.style;

    console.log("clicked", style);

    fetchData().then(() => {
      console.log("callback", style);
    });
  }

  render() {
    return (
      <button onClick={this.onClick}>{this.props.children}
        </button>
    );
  }
}

export default MyButton;
```

The second call to `console.log()` is attempting to access a synthetic event property from an asynchronous callback that doesn't run until the event handler completes, which causes the event to empty its properties. This results in a warning and an undefined value.

> **Note**
>
> The aim of this example is to illustrate how things can break when you write asynchronous code that interacts with events. Just don't do it!

In this section, you learned that events are pooled for performance reasons, which means that you should never access event objects in an asynchronous way.

Summary

This chapter introduced you to event handling in React. The key differentiator between React and other approaches to event handling is that handlers are declared in JSX markup. This makes tracking down which elements handle which events much simpler.

You learned that having multiple event handlers on a single element is a matter of adding new JSX properties. Next, you learned that it's a good idea to share event-handling functions that handle generic behavior. Context can be important for event handler functions if they need access to component properties or state. You learned about the various ways to bind event handler function context and parameter values. These include calling `bind()` and using higher-order event handler functions.

Then, you learned about inline event handler functions and their potential use, as well as how React actually binds a single DOM event handler to the document object. Synthetic events are abstractions that wrap native events; you learned why they're necessary and how they're pooled for efficient memory consumption.

In the next chapter, you'll learn how to create components that are reusable for a variety of purposes. Instead of writing new components for each use case that you encounter, you'll learn the skills necessary to refactor existing components so that they can be used in more than one context.

Further reading

Visit the following link for more information:

- *Handling Events* (`https://reactjs.org/docs/handling-events.html`).

6
Crafting Reusable Components

The focus of this chapter is to show you how to implement React components that serve more than just one purpose. After reading this chapter, you'll feel confident about how to compose application features.

The chapter starts with a brief look at HTML elements and how they work in terms of helping to implement features versus having a high level of utility. Then, you'll see the implementation of a monolithic component and discover the issues that it will cause down the road. The next section is devoted to re-implementing the monolithic component in such a way that the feature is composed of smaller components.

Finally, the chapter ends with a discussion of rendering trees of React components and gives you some tips on how to avoid introducing too much complexity as a result of decomposing components. I'll close the final section by reiterating the concept of high-level feature components versus utility components.

The following topics will be covered in this chapter:

- Reusable HTML elements
- The difficulty with monolithic components
- Refactoring component structures
- Render props
- Refactoring class components using hooks
- Rendering component trees
- Feature components and utility components

Technical requirements

You can find the code files for this chapter on GitHub at `https://github.com/PacktPublishing/React-and-React-Native-4th-Edition/tree/main/Chapter06`.

Reusable HTML elements

Let's think about HTML elements for a moment. Depending on the type of HTML element, it's either feature-centric or utility-centric. Utility-centric HTML elements are more reusable than feature-centric HTML elements. For example, consider the `<section>` element. This is a generic element that can be used just about anywhere, but its primary purpose is to compose the structural aspects of a feature: the outer shell of the feature and the inner sections of the feature. This is where the `<section>` element is most useful.

On the other side of the fence, you have elements such as `<p>`, `<span>`, and `<button>`. These elements provide a high level of utility because they're generic by design. You're supposed to use `<button>` elements whenever you have something that's clickable by the user, resulting in an action. This is a level lower than the concept of a feature.

While it's easy to talk about HTML elements that have a high level of utility versus those that are geared toward specific features, the discussion is more detailed when data is involved. HTML is static markup; React components combine static markup with data. The question is, how do you make sure that you're creating the right feature-centric and utility-centric components?

The aim of this chapter is to find out how to go from a monolithic React component that defines a feature to a smaller feature-centric component combined with utility components.

The difficulty with monolithic components

If you could implement just one component for any given feature, it would simplify your job. At the very least, there wouldn't be many components to maintain, and there wouldn't be many communication paths for data to flow through, because everything would be internal to the component.

However, this idea doesn't work for a number of reasons. Having monolithic feature components makes it difficult to coordinate any kind of team development effort. The bigger monolithic components become, the more difficult they are to refactor into something better later on.

There's also the problem of feature overlap and feature communication. Overlap happens because of similarities between features; it's unlikely that an application will have a set of features that are completely unique to one another. That would make the application very difficult to learn and use. Component communication essentially means that the state of something in one feature will impact the state of something in another feature. State is difficult to deal with, and even more so when there is a lot of state packaged up in a monolithic component.

The best way to learn how to avoid monolithic components is to experience one firsthand. You'll spend the remainder of this section implementing a monolithic component. In the following section, you'll see how this component can be refactored into something a little more sustainable.

The JSX markup

The monolithic component we're going to implement is a feature that lists articles. It's just for illustrative purposes, so we don't want to go overboard on the size of the component. It'll be simple, yet monolithic. The user can add new items to the list, toggle the summary of items in the list, and remove items from the list.

Here is the `render()` method of the component:

```
render() {
  const { articles, title, summary } = this.state;

  return (
    <section>
      <header>
        <h1>Articles</h1>
        <input
          placeholder="Title"
```

```
      value={title}
      onChange={this.onChangeTitle}
    />
    <input
      placeholder="Summary"
      value={summary}
      onChange={this.onChangeSummary}
    />
    <button onClick={this.onClickAdd}>Add</button>
  </header>
  <article>
    <ul>
      {articles.map((i) => (
        <li key={i.id}>
          <a
            href={'#${i.id}'}
            title="Toggle Summary"
            onClick={this.onClickToggle.bind
              (null, i.id)}
          >
            {i.title}
          </a>
           
          <a
            href={'#${i.id}'}
            title="Remove"
            onClick={this.onClickRemove.bind
              (null, i.id)}
          >
            &#10007;
          </a>
          <p style={{ display: i.display
            }}>{i.summary}</p>
        </li>
      ))}
    </ul>
```

```
        </article>
      </section>
    );
  }
```

This is definitely more JSX than is necessary in one place. We'll improve on this in the following section, but for now, let's implement the initial state for this component.

I strongly encourage you to download the companion code for this book from `https://github.com/PacktPublishing/React-and-React-Native-4th-Edition`. I can break apart the component code so that I can explain it on these pages. However, it's an easier learning experience if you can see the code modules in their entirety, in addition to running them.

Initial state

Now, let's look at the initial state of this component:

```
state = {
  articles: [
    {
      id: id.next(),
      title: "Article 1",
      summary: "Article 1 Summary",
      display: "none",
    },
    {
      id: id.next(),
      title: "Article 2",
      summary: "Article 2 Summary",
      display: "none",
    },
    {
      id: id.next(),
      title: "Article 3",
      summary: "Article 3 Summary",
      display: "none",
    },
    {
```

```
      id: id.next(),
      title: "Article 4",
      summary: "Article 4 Summary",
      display: "none",
    },
  ],
  title: "",
  summary: "",
};
```

The state consists of an array of `articles`, a `title` string, and a `summary` string. Each article object in the `articles` array has several string fields to help render the article and an `id` field, which is a number. The number is generated by `id.next()`.

Let's take a look at how this works:

```
const id = (function* () {
  let i = 1;
  while (true) {
    yield i;
    i += 1;
  }
})();
```

The `id` constant is a generator. It is created by defining an inline generator function and calling it right away. This generator will yield numbers infinitely. So, calling `id.next()` the first time returns `1`, the next is `2`, and so on. This simple utility will come in handy when it's time to add new articles and we need a new unique ID.

Event handler implementation

At this point, you have the initial state and the JSX of the component. Now, it's time to implement the event handlers:

```
onChangeTitle = (e) => {
  this.setState({ title: e.target.value });
};

onChangeSummary = (e) => {
  this.setState({ summary: e.target.value });
};
```

The `onChangeTitle()` and `onChangeSummary()` methods use `setState()` to update the `title` and `summary` state values, respectively. The new values come from the `target.value` property of the `event` argument, which is the value that the user types into the text input:

```
onClickAdd = () => {
  this.setState((state) => ({
    articles: [
      ...state.articles,
      {
        id: id.next(),
        title: state.title,
        summary: state.summary,
        display: "none",
      },
    ],
    title: "",
    summary: "",
  }));
};
```

The `onClickAdd()` method adds a new article to the `articles` state. This state value is an array. We use the spread operator to build a new array from the existing array (`[...state.articles]`), and the new object gets added to the end of the new array. The reason we're building a new array and passing it to `setState()` is so that there are no surprises. In other words, we're treating state values as immutable so that other code that updates the same state doesn't accidentally cause problems:

```
onClickRemove = (id) => {
  this.setState((state) => ({
    ...state,
    articles: state.articles.filter((article) =>
      article.id !== id),
  }));
};
```

The `onClickRemove()` method removes the article with the given ID from the `articles` state. It does this by calling `filter()` on the array, which returns a new array, so the operation is immutable.

The filter removes the object with the given ID:

```
onClickToggle = (id) => {
  this.setState((state) => {
    const articles = [...state.articles];
    const index = articles.findIndex(
      (article) => article.id === id
    );

    articles[index] = {
      ...articles[index],
      display: articles[index].display ? "" : "none",
    };

    return { ...state, articles };
  });
};
```

The `onClickToggle()` method toggles the visibility of the article with the given ID. We carry out two immutable operations in this method. First, we build a new `articles` array from `state.articles`. Then, based on the index of the given ID, we replace the article object at the index with a new object. We use the object spread operator to fill in the properties (`{...articles[index]}`), and then the `display` property value is toggled based on the existing display value.

Here's a screenshot of the output rendered:

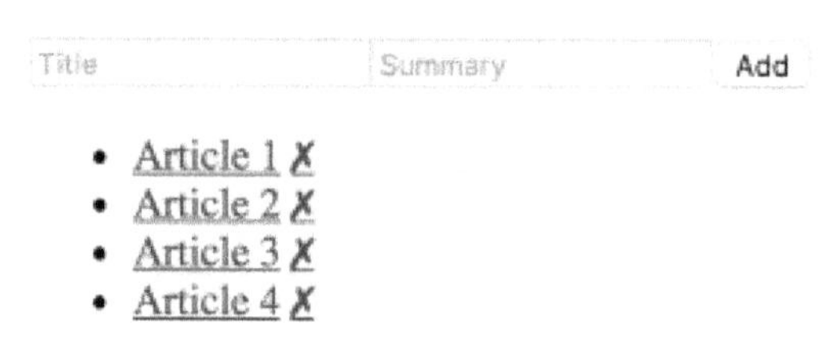

Figure 6.1 – Rendered articles

At this point, we have a component that does everything that we need our feature to do. However, it's monolithic and difficult to maintain. Imagine if we had other places in our app that use the same pieces of `MyFeature`? They have to re-invent them because they cannot be shared. In the following section, we'll work on breaking down `MyFeature` into smaller reusable components.

Refactoring component structures

You have a monolithic feature component—now what? Let's make it better.

In this section, you'll learn how to take the feature component that you just implemented in the preceding section and split it into more maintainable components. You'll start with the JSX, as this is probably the best refactor starting point. Then, you'll implement new components for the feature.

Next, you'll make these new components functional, instead of class-based. Finally, you'll learn how to use render props to reduce the number of direct component dependencies in your application, and how to remove classes entirely by using hooks to manage state within functional components.

Starting with the JSX

The JSX of any monolithic component is the best starting point for figuring out how to refactor it into smaller components. Let's visualize the structure of the component that we're currently refactoring:

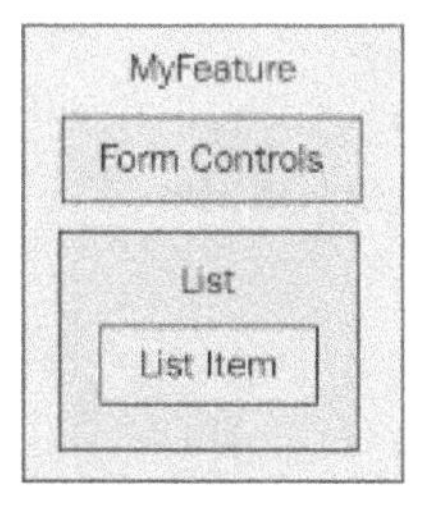

Figure 6.2 – Visualization of the JSX that makes up a React component

The top part of the JSX is the form controls, so this could easily become its own component:

```
<header>
  <h1>Articles</h1>
  <input
    placeholder="Title"
    value={title}
    onChange={this.onChangeTitle}
  />
  <input
    placeholder="Summary"
    value={summary}
    onChange={this.onChangeSummary}
  />
  <button onClick={this.onClickAdd}>Add</button>
</header>
```

Next, you have the list of articles:

```
<ul>
  {articles.map((i) => (
    <li key={i.id}>
      <a
        href={'#${i.id}'}
        title="Toggle Summary"
        onClick={this.onClickToggle.bind(null, i.id)}
      >
        {i.title}
      </a>
       
      <a
        href={'#${i.id}'}
        title="Remove"
        onClick={this.onClickRemove.bind(null, i.id)}
      >
        &#10007;
      </a>
```

```
      <p style={{ display: i.display }}>{i.summary}</p>
    </li>
  ))}
</ul>
```

Within this list, there's potential for an article component, which would be everything in the `<li>` tag. Let's try building this next.

Implementing an article list component

Here's what the article list component implementation looks like:

```
class ArticleList extends React.Component {
  render() {
    const { articles, onClickToggle, onClickRemove } =
      this.props;

    return (
      <ul>
        {articles.map((article) => (
          <li key={article.id}>
            <a
              href={'#${article.id}'}
              title="Toggle Summary"
              onClick={onClickToggle.bind(null,
                article.id)}
            >
              {article.title}
            </a>
             
            <a
              href={'#${article.id}'}
              title="Remove"
              onClick={onClickRemove.bind(null,
                article.id)}
            >
              &#10007;
```

```
            </a>
            <p style={{ display: article.display }}>
              {article.summary}
            </p>
          </li>
        ))}
      </ul>
    );
}
```

We're taking the relevant JSX out of the monolithic component and putting it here. Now, let's see what the feature component of JSX looks like:

```
render() {
  const { articles, title, summary } = this.state;

  return (
    <section>
      <header>
        <h1>Articles</h1>
        <input
          placeholder="Title"
          value={title}
          onChange={this.onChangeTitle}
        />
        <input
          placeholder="Summary"
          value={summary}
          onChange={this.onChangeSummary}
        />
        <button onClick={this.onClickAdd}>Add</button>
      </header>
      <ArticleList
        articles={articles}
        onClickToggle={this.onClickToggle}
        onClickRemove={this.onClickRemove}
      />
```

```
      </section>
    );
  }
```

The list of articles is now rendered by the `<ArticleList>` component. The list of articles to render is passed to this component as a property along with two of the event handlers.

> **Note**
> Why are we passing event handlers to a child component? The reason is so that the `ArticleList` component doesn't have to worry about the state or how the state changes. All it cares about is rendering content and making sure the appropriate event callbacks are hooked up to the appropriate DOM elements. This is a container component concept that I'll expand upon later in this chapter.

Now that we have an `ArticleList` component, let's see whether we can further break it down into smaller reusable components.

Implementing an article item component

After implementing the article list component, you might decide that it's a good idea to break this component.

Another way to look at it is this is, if it turns out that we don't actually need the item as its own component, this new component doesn't introduce much indirection or complexity. Without further ado, here's the article item component:

```
import * as React from "react";

class ArticleItem extends React.Component {
  render() {
    const { article, onClickToggle, onClickRemove } =
      this.props;

    return (
      <li>
        <a
          href={'#{article.id}'}
          title="Toggle Summary"
```

```
          onClick={onClickToggle.bind(null, article.id)}
        >
          {article.title}
        </a>
         
        <a
          href={'#{article.id}'}
          title="Remove"
          onClick={onClickRemove.bind(null, article.id)}
        >
          &#10007;
        </a>
        <p style={{ display: article.display }}>
          {article.summary}
        </p>
      </li>
    );
  }
}

export default ArticleItem;
```

Here's the new `ArticleItem` component being rendered by the `ArticleList` component:

```
import * as React from "react";
import ArticleItem from "./ArticleItem";

class ArticleList extends React.Component {
  render() {
    const { articles, onClickToggle, onClickRemove } =
      this.props;

    return (
      <ul>
        {articles.map((i) => (
          <ArticleItem
```

```
            key={i.id}
            article={i}
            onClickToggle={onClickToggle}
            onClickRemove={onClickRemove}
          />
        ))}
      </ul>
    );
  }
}

export default ArticleList;
```

Do you see how this list just maps the list of articles? What if you wanted to implement another article list that does some filtering too? If so, it's beneficial to have a reusable `ArticleItem` component. Next, we'll move the add article markup into its own component.

Implementing an add article component

Now that we're done with the article list, it's time to think about the form controls used to add a new article. Let's implement a component for this aspect of the feature:

```
import * as React from "react";

class AddArticle extends React.Component {
  render() {
    const {
      name,
      title,
      summary,
      onChangeTitle,
      onChangeSummary,
      onClickAdd,
    } = this.props;

    return (
      <section>
```

```
        <h1>{name}</h1>
        <input
          placeholder="Title"
          value={title}
          onChange={onChangeTitle}
        />
        <input
          placeholder="Summary"
          value={summary}
          onChange={onChangeSummary}
        />
        <button onClick={onClickAdd}>Add</button>
      </section>
    );
  }
}

export default AddArticle;
```

Now, our feature component only needs to render `<AddArticle>` and `<ArticleList>` components:

```
render() {
  const { articles, title, summary } = this.state;

  return (
    <section>
      <AddArticle
        name="Articles"
        title={title}
        summary={summary}
        onChangeTitle={this.onChangeTitle}
        onChangeSummary={this.onChangeSummary}
        onClickAdd={this.onClickAdd}
      />

      <ArticleList
```

```
        articles={articles}
        onClickToggle={this.onClickToggle}
        onClickRemove={this.onClickRemove}
      />
    </section>
  );
}
```

The focus of this component is on the feature data, while it defers to other components for rendering UI elements. Several components that we've created while refactoring `MyFeature` are classes, and they don't need to be. Let's make them simple functions instead.

Making components functional

While implementing these new components, you may have noticed that they don't have any responsibilities other than rendering JSX using property values. These components are good candidates for pure function components. Whenever you come across components that only use property values, it's a good idea to make them functional. For one thing, it makes it explicit that the component doesn't rely on any state or life cycle methods. It's also more efficient because React doesn't perform as much work when it detects that a component is a function.

Here is the functional version of the `ArticleList` component:

```
import ArticleItem from "./ArticleItem";

function ArticleList({ articles, onClickToggle,
  onClickRemove }) {
  return (
    <ul>
      {articles.map((i) => (
        <ArticleItem
          key={i.id}
          article={i}
          onClickToggle={onClickToggle}
          onClickRemove={onClickRemove}
        />
      ))}
```

```
    </ul>
  );
}

export default ArticleList;
```

Here is the functional version of the `ArticleItem` component:

```
function ArticleItem({ article, onClickToggle,
  onClickRemove }) {
  return (
    <li>
      <a
        href={'#${article.id}'}
        title="Toggle Summary"
        onClick={onClickToggle.bind(null, article.id)}
      >
        {article.title}
      </a>
       
      <a
        href={'#${article.id}'}
        title="Remove"
        onClick={onClickRemove.bind(null, article.id)}
      >
        &#10007;
      </a>
      <p style={{ display: article.display
        }}>{article.summary}</p>
    </li>
  );
}

export default ArticleItem;
```

Here is the functional version of the `AddArticle` component:

```
function AddArticle({
  name,
  title,
  summary,
  onChangeTitle,
  onChangeSummary,
  onClickAdd,
}) {
  return (
    <section>
      <h1>{name}</h1>
      <input
        placeholder="Title"
        value={title}
        onChange={onChangeTitle}
      />
      <input
        placeholder="Summary"
        value={summary}
        onChange={onChangeSummary}
      />
      <button onClick={onClickAdd}>Add</button>
    </section>
  );
}

export default AddArticle;
```

Another added benefit of making components functional is that there's less opportunity to introduce unnecessary methods or other data.

In this section, you learned about using JSX as the basis for refactoring larger components into smaller, more reusable ones. This leads to more components, but they're smaller, more focused, and reusable. In the next section, we'll look at how render props make it possible to pass components around as properties instead of directly importing them as dependencies.

Render props

Imagine implementing a feature that is composed of several smaller components, like what you've been working on in this chapter. The `MyFeature` component depends on `ArticleList` and `AddArticle`. Now, imagine using `MyFeature` in different parts of your application where it makes sense to use a different implementation of `ArticleList` or `AddArticle`. The fundamental challenge is substituting one component for another.

Render props are a nice way to address this challenge. The idea is that you pass a property to your component whose value is a function that returns a component to render. This way, instead of having the feature component directly depend on its child components, you can configure them as you like; they pass them in as render prop values.

> **Note**
>
> Render props aren't a React 16 feature. It's a technique whose popularity increase coincided with the release of React 16. It's an officially recognized way to deal with dependency and substitution problems. You can read more about render props at `https://reactjs.org/docs/render-props.html`.

Let's look at an example. Instead of having `MyFeature` directly depend on `AddArticle` and `ArticleList`, you can pass them as render props. Here's what the `render()` method of `MyFeature` looks like when it's using render props to fill in the holes where `add` used to be:

```
render() {
  const { articles, title, summary } = this.state;
  const {
    props: { addArticle, articleList },
    onClickAdd,
    onClickToggle,
    onClickRemove,
    onChangeTitle,
    onChangeSummary,
  } = this;

  return (
    <section>
      {addArticle({
```

```
        title,
        summary,
        onChangeTitle,
        onChangeSummary,
        onClickAdd,
      })}
      {articleList({ articles, onClickToggle, onClickRemove
        })}
    </section>
  );
}
```

The `addArticle()` and `articleList()` functions are called with the same property values that would have been passed to `<AddArticle>` and `<ArticleList>`, respectively. The difference now is that this module no longer imports `AddArticle` or `ArticleList` as dependencies.

Now, let's take a look at the `index.js` file where `<MyFeature>` is rendered:

```
const root =
  ReactDOM.createRoot(document.getElementById("root"));
root.render(
  <MyFeature
    addArticle={({
      title,
      summary,
      onChangeTitle,
      onChangeSummary,
      onClickAdd,
    }) => (
      <AddArticle
        name="Articles"
        title={title}
        summary={summary}
        onChangeTitle={onChangeTitle}
        onChangeSummary={onChangeSummary}
        onClickAdd={onClickAdd}
      />
```

```
    )}
    articleList={({ articles, onClickToggle, onClickRemove
      }) => (
      <ArticleList
        articles={articles}
        onClickToggle={onClickToggle}
        onClickRemove={onClickRemove}
      />
    )}
  />
);
```

There's a lot more going on here now than there was when it was just `<MyFeature>` being rendered. Let's break down why that is. Here is where you pass the `addArticle` and `articleList` render props. These prop values are functions that accept argument values from `MyComponent`. For example, the `onClickToggle()` function comes from `MyFeature` and is used to change the state of that component. You can use the render prop function to pass this to the component that will be rendered, along with any other values. The return value of these functions is what is ultimately rendered.

In this section, you learned that by passing render property values – functions that render JSX markup – you can avoid hardcoding dependencies in places where you might want to share functionality. Passing a different property value to a component is usually easier than changing the dependencies used by a given module. In the final section of this chapter, we'll refactor the `MyFeature` component into a functional component that uses hooks for state management.

Refactoring class components using hooks

Prior to the addition of hooks to React, we would often end up using class-based components just because the component had state data to maintain. Hooks exist so that you can implement React components using regular functions and still have access to the React APIs that you used to access through class attributes and methods. In this section, we'll rewrite the `MyFeature` component so that it's a function and it uses the `useState()` hook.

First, let's take a look at the functional version of `MyFeature`:

```
function MyFeature({ addArticle, articleList }) {
  const [articles, setArticles] = React.useState([
    {
      id: id.next(),
      title: "Article 1",
      summary: "Article 1 Summary",
      display: "none",
    },
    ...
  ]);

  const [title, setTitle] = React.useState("");
  const [summary, setSummary] = React.useState("");

  function onChangeTitle(e) {
    setTitle(e.target.value);
  }

  function onChangeSummary(e) {
    setSummary(e.target.value);
  }

  function onClickAdd() {
    setArticles([
      ...articles,
      {
        id: id.next(),
        title: title,
        summary: summary,
        display: "none",
      },
    ]);
    setTitle("");
    setSummary("");
  }
```

```
  function onClickRemove(id) {
    setArticles(articles.filter((article) => article.id
      !== id));
  }

  function onClickToggle(id) {
    const index = articles.findIndex((article) =>
      article.id === id);
    const updatedArticles = [...articles];

    updatedArticles[index] = {
      ...articles[index],
      display: articles[index].display ? "" : "none",
    };

    setArticles(updatedArticles);
  }

  return (
    <section>
      {addArticle({
        title,
        summary,
        onChangeTitle,
        onChangeSummary,
        onClickAdd,
      })}
      {articleList({ articles, onClickToggle,
        onClickRemove })}
    </section>
  );
}
```

Even though we've completely changed the implementation of `MyFeature`, none of the other utility components, such as `AddArticle` or `ArticleList`, require any changes. Now, let's take a closer look at what was changed, starting with the component declaration:

```
function MyFeature({ addArticle, articleList }) {
  ...
}
```

Now, `MyFeature` is a function that takes two properties (`addArticle` and `articleList`) as arguments. Next, let's look at how state is initialized in this function:

```
const [articles, setArticles] = React.useState([
  {
    id: id.next(),
    title: "Article 1",
    summary: "Article 1 Summary",
    display: "none",
  },
  {
    id: id.next(),
    title: "Article 2",
    summary: "Article 2 Summary",
    display: "none",
  },
  {
    id: id.next(),
    title: "Article 3",
    summary: "Article 3 Summary",
    display: "none",
  },
  {
    id: id.next(),
    title: "Article 4",
    summary: "Article 4 Summary",
    display: "none",
  },
]);
```

```
const [title, setTitle] = React.useState("");
const [summary, setSummary] = React.useState("");
```

Now, instead of assigning the pieces of state that our component needs to a `state` property on a class, we're using the `useState()` hook to initialize our state values and state setter functions. One immediate benefit of this approach is that the state values are now accessible throughout the function scope. We no longer need to access state values via `this.state`.

Next, let's look at the event handler implementations:

```
function onChangeTitle(e) {
  setTitle(e.target.value);
}

function onChangeSummary(e) {
  setSummary(e.target.value);
}

function onClickAdd() {
  setArticles([
    ...articles,
    {
      id: id.next(),
      title: title,
      summary: summary,
      display: "none",
    },
  ]);
  setTitle("");
  setSummary("");
}

function onClickRemove(id) {
  setArticles(articles.filter((article) => article.id
    !== id));
}
```

```
function onClickToggle(id) {
  const index = articles.findIndex((article) => article.id
    === id);
  const updatedArticles = [...articles];

  updatedArticles[index] = {
    ...articles[index],
    display: articles[index].display ? "" : "none",
  };

  setArticles(updatedArticles);
}
```

Now, instead of using `this.setState()` to update any values, we can just use the setter functions. For example, `setArticles()` updates the `articles` state. In cases where updating the state depends on the previous state value, we can simply access the previous value directly. For example, in the `onClickToggle()` handler, we need access to the `articles` array before we can update it. The `articles` constant is available to us to read the current state value, which leads to simpler code; we no longer need to pass a callback function to `setState()`.

The callbacks are now functions nested inside the `MyFeature` function, instead of class methods. The functions are named, so no readability is lost. Also, there's no scope to worry about since everything, including state values, is within the larger component function scope.

This section showed you how to take an existing class component that has state and refactor it into a functional component with state. The `useState()` hook leads to a simplified state management code. In the following section, we'll look at the concept of component trees.

Rendering component trees

Let's take a moment to reflect on what we've accomplished so far in this chapter. The feature component that was once monolithic ended up focusing almost entirely on the state data. It handled the initial state and handled transforming the state, and it would handle network requests that fetch state, if there were any. This is a typical container component in a React application, and it's the starting point for data.

The new components that you implemented, to better compose the feature, were the recipients of this data. The difference between these components and their container is that they only care about the properties that are passed into them at the time they're rendered. In other words, they only care about data snapshots at a particular point in time. From here, these components might pass the property data into their own child components as properties. The generic pattern for composing React components is as follows:

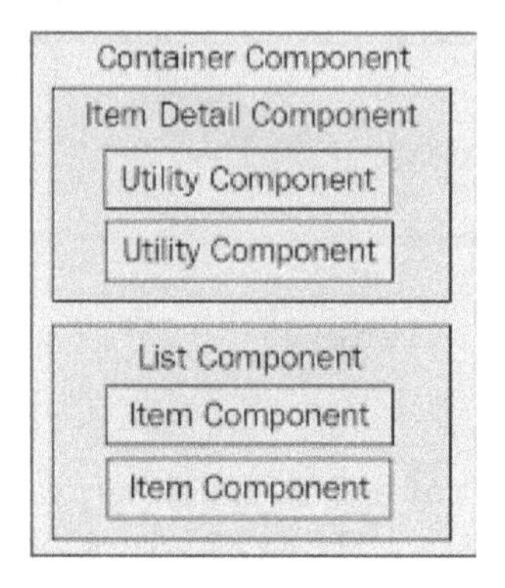

Figure 6.3 – A pattern for composing larger React components from smaller components

The container component will typically contain one direct child. In this diagram, you can see that the container has either an item detail component or a list component. Of course, there will be variations in these two categories, as every application is different. This generic pattern has three levels of component composition. Data flows in one direction from the container all the way down to the utility components.

Once you add more than three layers, the application architecture becomes difficult to comprehend. There will be the odd case where you'll need to add four layers of React components but, as a rule of thumb, you should avoid this.

Feature components and utility components

In the monolithic component example, you started off with a single component that was entirely focused on a feature. This means that the component has very little utility elsewhere in the application.

The reason for this is that top-level components deal with the application state. Stateful components are difficult to use in any other context. As you refactored the monolithic feature component, you created new components that moved further away from the data. The general rule is that the further your components move from stateful data, the more utility they have because their property values could be passed in from anywhere in the application.

Summary

This chapter was about avoiding a monolithic component design. However, monoliths are often a necessary starting point in the design of any React component.

You began by learning about how the different HTML elements have varying degrees of utility. Next, you learned about the issues with monolithic React components and walked through the implementation of a monolithic component.

Then, you spent several sections learning how to refactor the monolithic component into a more sustainable design. From this exercise, you learned that container components should only have to think in terms of handling state, while smaller components have more utility because their property values can be passed from anywhere. You also learned that you could use render props for better control over component dependencies and substitution.

In the next chapter, you'll learn about the React component life cycle. This is an especially relevant topic for implementing container components.

Further reading

Visit the following links for more information:

- **Render props**: `https://reactjs.org/docs/render-props.html`
- **Components and props**: `https://reactjs.org/docs/components-and-props.html`

7
The React Component Life Cycle

The goal of this chapter is for you to learn about the life cycle of React components and how to write code that responds to life cycle events. You'll learn why components need a life cycle in the first place. Then, you'll implement several components that initialize their properties and state using these methods.

Next, you'll learn about how to optimize the rendering efficiency of your components by avoiding rendering when it isn't necessary. Then, you'll see how to encapsulate the imperative code in React components and how to clean up when components are unmounted. Finally, you'll learn how to capture and handle errors using React life cycle methods.

Here are the sections we'll cover in this chapter:

- Why components need a life cycle
- Initializing properties and state
- Optimizing rendering efficiency
- Rendering imperative components
- Cleaning up after components
- Containing errors with error boundaries

Technical requirements

You can find the code files for this chapter on GitHub at `https://github.com/PacktPublishing/React-and-React-Native-4th-Edition/tree/main/Chapter07`.

Why components need a life cycle

React components go through a life cycle. In fact, the `render()` method that you've implemented in your components so far in this book is actually a life cycle method. Rendering is just one life cycle event in a React component.

For example, there are life cycle events for when the component is mounted to the DOM, when the component is updated, and so on. Life cycle events are yet another moving part, so you'll want to keep them to a minimum. As you'll learn in this chapter, some components do need to respond to life cycle events to perform initialization, render heuristics, clean up after the component when it's unmounted from the DOM, or handle errors thrown by the component.

The following diagram gives you an idea of how a component flows through its life cycle, calling the corresponding methods in turn:

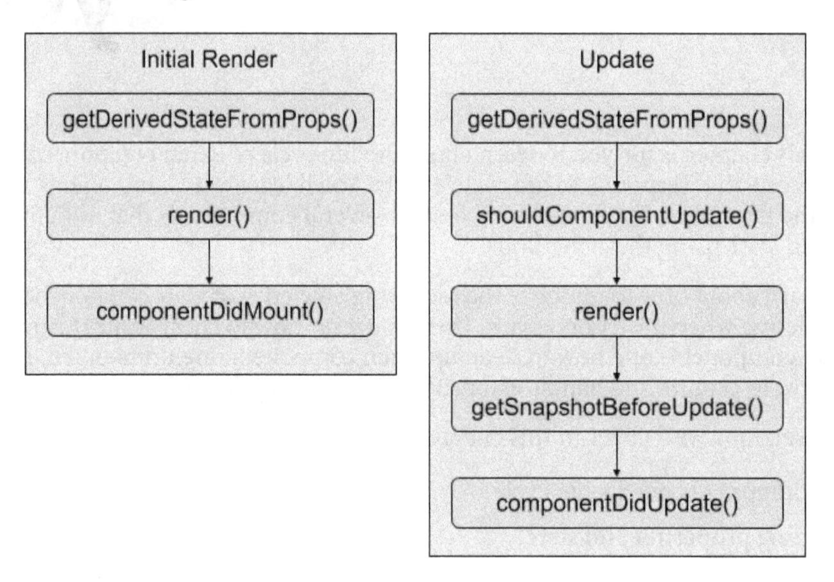

Figure 7.1 – The functions used in the life cycle of React components

These are the two main life cycle flows of a React component. The first happens when the component is initially rendered. The second happens whenever the component is updated. Here's a rough overview of each of the methods:

- `getDerivedStateFromProps()`: This method allows you to update the state of the component based on the property values of the component. This method is called when the component is initially rendered and when it receives new property values.
- `render()`: This returns the content to be rendered by the component. This is called when the component is first mounted to the DOM, when it receives new property values, and when `setState()` is called.
- `componentDidMount()`: This is called after the component is mounted to the DOM. This is where you can perform component initialization work, such as fetching data.
- `shouldComponentUpdate()`: You can use this method to compare new states or props with current states or props. Then, you can return `false` if there's no need to re-render the component. This method is used to make your components more efficient.
- `getSnapshotBeforeUpdate()`: This method lets you perform operations directly on the DOM elements of your component before they're actually committed to the DOM. The difference between this method and `render()` is that `getSnapshotBeforeUpdate()` isn't asynchronous. With `render()`, there's a good chance that the DOM structure could change between when it's called and when the changes are actually made in the DOM.
- `componentDidUpdate()`: This is called when the component is updated. It's rare that you'll have to use this method.

The other life cycle method that isn't included in this diagram is `componentWillUnmount()`. This is the only life cycle method that's called when a component is about to be removed. We'll see an example of how to use this method at the end of this chapter. On that note, let's get coding.

Initializing properties and state

In this section, you'll see how to implement the initialization code in React components. This involves using life cycle methods that are called when the component is first created. First, you'll implement a basic example that sets the component up with data from the API. Then, you'll see how the state can be initialized from properties, and also how the state can be updated as properties change.

Fetching component data

When your components are initialized, you'll want to populate their state or properties. Otherwise, the component won't have anything to render other than its skeleton markup. For instance, let's say you want to render the following user list component:

```
const ErrorMessage = ({ error }) =>
  error ? <strong>{error}</strong> : null;

const LoadingMessage = ({ loading }) =>
  loading ? <em>{loading}</em> : null;

function UserList({ error, loading, users }) {
  return (
    <section>
      <ErrorMessage error={error} />
      <LoadingMessage loading={loading} />
      <ul>
        {users.map((user) => (
          <li key={user.id}>{user.name}</li>
        ))}
      </ul>
    </section>
  );
}
```

There are three pieces of data that this JSX relies on, as follows:

- `loading`: This message is displayed while fetching API data.
- `error`: This message is displayed if something goes wrong.
- `users`: Data that's fetched from the API.

There are two helper components being used here: `ErrorMessage` and `LoadingMessage`. They're used to format the error and the loading states, respectively. If `error` or `loading` is null, nothing is rendered. Otherwise, an error or loading message is rendered by the respective component.

How should we go about making the API call and using the response to populate the `users` collection? The answer is to use a container component that makes the API call and then renders the `UserList` component:

```
import { users } from "./api";
import UserList from "./UserList";

class UserListContainer extends React.Component {
  state = {
    error: null,
    loading: "loading...",
    users: [],
  };

  componentDidMount() {
    users().then(
      (result) => {
        this.setState({
          loading: null,
          error: null,
          users: result.users,
        });
      },
      (error) => {
        this.setState({ loading: null, error });
      }
    );
  }

  render() {
    return <UserList {...this.state} />;
  }
}
```

Let's take a look at the `render()` method. Its job is to render the `<UserList>` component, passing in `this.state` as properties. The actual API call happens in the `componentDidMount()` method. This method is called after the component is mounted into the DOM.

> **Note**
>
> Due to the naming of `componentDidMount()`, React newcomers think that it's bad to wait until the component is mounted to the DOM before issuing requests for component data. In other words, the user experience might suffer if React has to perform a lot of work before the request is even sent. In reality, fetching data is an asynchronous task, and initiating it before or after `render()` makes no real difference as far as your application is concerned. You can read more about this at `https://reactjs.org/blog/2018/03/27/update-on-async-rendering.html`.

Once the API call returns with data, the `users` collection is populated, causing the `UserList` to re-render itself, only this time, it has the data it needs. Let's take a look at the `users()` mock API function call being used here:

```
export function users(fail) {
  return new Promise((resolve, reject) => {
    setTimeout(() => {
      if (fail) {
        reject("epic fail");
      } else {
        resolve({
          users: [
            { id: 0, name: "First" },
            { id: 1, name: "Second" },
            { id: 2, name: "Third" }
          ]
        });
      }
    }, 2000);
  });
}
```

It returns a promise that's resolved with an array after 2 seconds. Promises are a good tool for mocking things such as API calls because they enable you to use more than HTTP calls as a data source in your React components. For example, you might be reading from a local file or using a library that returns promises that resolve data from various sources.

Here's what the `UserList` component renders when the loading state is a string, and the `users` state is an empty array:

loading...

Figure 7.2 – The loading screen when the users state is empty

Here's what it renders when `loading` is null and `users` is non-empty:

- First
- Second
- Third

Figure 7.3 – Rendering a list when the users state is populated

I want to reiterate the separation of responsibilities between the `UserListContainer` and `UserList` components. Because the container component handles the life cycle management and the actual API communication, you can create a generic user list component. In fact, it's a functional component that doesn't require any state, which means you can reuse it in other container components throughout your application.

Now that we've seen how to set the state of a component using fetched API data, let's figure out how to set the state of a component using property values that are passed to it.

Initializing state with properties

The preceding example showed you how to initialize the state of a container component by making an API call in the `componentDidMount()` life cycle method. However, the only populated part of the component state was the `users` collection. You might want to populate other pieces of state that don't come from API endpoints.

For example, the `error` and `loading` state messages have default values set when the state is initialized. This is great, but what if the code that is rendering `UserListContainer` wants to use a different loading message? You can achieve this by allowing properties to override the default state. Let's build on the `UserListContainer` component:

```
import { users } from "./api";
import UserList from "./UserList";

class UserListContainer extends React.Component {
  state = {
    error: null,
    users: [],
  };
```

```
  componentDidMount() {
    users().then(
      (result) => {
        this.setState({ error: null, users: result.users
          });
      },
      (error) => {
        this.setState({ loading: null, error });
      }
    );
  }

  render() {
    return <UserList {...this.state} />;
  }

  static getDerivedStateFromProps(props, state) {
    return {
      ...state,
      loading: state.users.length === 0 ? props.loading :
        null,
    };
  }
}

UserListContainer.defaultProps = {
  loading: "loading...",
};
```

The `loading` property no longer has a default string value. Instead, `defaultProps` provides default values for properties. The new life cycle method is `getDerivedStateFromProps()`. It uses the `loading` property to set the loading state. Since the `loading` property has a default value, it's safe to just change the state. The method is called before the component mounts and on subsequent re-renders of the component.

> **Note**
>
> This method is static because of internal changes in React 16. The expectation is that this method behaves like a pure function and has no side effects. If this method were an instance method, you would have access to the component context and side effects would be commonplace.

The challenge with this method is that it's called on the initial render and subsequent re-renders. Prior to React 16, you could use the `componentWillMount()` method for code that you only wanted to run prior to the initial render. In this example, you have to check whether there are values in the `users` collection before setting the `loading` state to `null` – you don't know whether this is the initial render or the fortieth render.

Let's see how we can pass state data to `UserListContainer` now:

```
import UserListContainer from "./UserListContainer";

const root =
  ReactDOM.createRoot(document.getElementById("root"));
root.render(
  <UserListContainer loading="playing the waiting game..."
    />
);
```

Here's what the initial loading message looks like when `UserList` is first rendered:

playing the waiting game...

Figure 7.4 – Displaying the loading message

Just because the component has state doesn't mean that you can't allow for customization. Next, you'll learn a variation of this concept—updating the component state with properties.

Updating state with properties

You've seen how the `componentDidMount()` and `getDerivedStateFromProps()` life cycle methods help get your components the data they need. There's one more scenario that you need to consider—re-rendering the component container.

Let's take a look at a simple button component that tracks the number of times it's been clicked:

```
function MyButton({ clicks, disabled, text, onClick }) {
  return (
    <section>
      <p>{clicks} clicks</p>
      <button disabled={disabled} onClick={onClick}>
        {text}
      </button>
    </section>
  );
}
```

Now, let's implement a container component for this feature:

```
import MyButton from "./MyButton";

class MyFeature extends React.Component {
  state = {
    clicks: 0,
    disabled: false,
    text: "",
  };

  onClick = () => {
    this.setState((state) => ({
      ...state,
      clicks: state.clicks + 1,
    }));
  };

  render() {
    return <MyButton onClick={this.onClick} {...this.state}
      />;
  }

  static getDerivedStateFromProps({ disabled, text },
```

```
    state) {
    return { ...state, disabled, text };
  }
}

MyFeature.defaultProps = {
  text: "A Button",
};
```

The same approach that we used for initializing the state with properties is being used here. The `getDerivedStateFromProps()` method is called before every render and is where you can use prop values to figure out whether and how the component state should be updated. Let's see how to re-render this component and whether or not the state behaves as expected:

```
import MyFeature from "./MyFeature";

let disabled = true;
const root =
  ReactDOM.createRoot(document.getElementById("root"));

function render() {
  disabled = !disabled;

  root.render(<MyFeature {...{ disabled }} />);
}

setInterval(render, 3000);
render();
```

Sure enough, everything goes as planned. Whenever the button is clicked, the click counter is updated. `<MyFeature>` is re-rendered every 3 seconds, toggling the `disabled` state of the button. When the button is re-enabled and clicking resumes, the counter continues from where it left off.

Here is what the `MyButton` component looks like when it's first rendered:

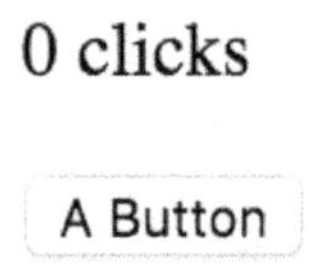

0 clicks

A Button

Figure 7.5 – The button hasn't been clicked yet

Here's what it looks like after it has been clicked a few times and the button has moved into a disabled state:

9 clicks

A Button

Figure 7.6 – Showing the number of times the button has been clicked

In this section, you learned about initializing property and state values in your components by using different life cycle methods. Without these methods, you would have a hard time ensuring that your components have the data that they need when they need it. In the next section, we'll consider different ways to optimize the efficiency of our components using life cycle methods.

Optimizing rendering efficiency

The next life cycle method you're going to learn about is used to implement heuristics that improve component rendering performance. You'll see that if the state of a component hasn't changed, then there's no need to render. Then, you'll implement a component that uses specific metadata from the API to determine whether or not the component needs to be re-rendered.

To render or not to render

The `shouldComponentUpdate()` life cycle method is used to determine whether or not the component will render when asked to. For example, if this method were implemented and returned `false`, the entire life cycle of the component would short-circuit, and no render would happen. This can be an important check to have in place if the component is rendering a lot of data and is re-rendered frequently. The trick is knowing whether or not the component state has changed.

Let's take a look at a simple list component:

```
function referenceEquality(arr1, arr2) {
  return arr1 === arr2;
}

function valueEquality(arr1, arr2) {
  for (let i = 0; i < arr1.length; i++) {
    if (arr1[i] !== arr2[i]) {
      return false;
    }
  }
  return true;
}

class MyList extends React.Component {
  state = {
    items: new Array(5000).fill(null).map((v, i) => i),
  };

  shouldComponentUpdate(props, state) {
    if (!referenceEquality(this.state.items, state.items))
    {
      return !valueEquality(this.state.items, state.items);
    }

    return false;
  }

  render() {
    return (
      <ul>
        {this.state.items.map((item) => (
          <li key={item}>{item}</li>
        ))}
      </ul>
```

```
    );
  }
}
```

The items state is initialized to an array with 5,000 items in it. This is a fairly large collection, so you don't want the virtual DOM inside React to constantly diff this list. The virtual DOM is efficient at what it does but not nearly as efficient as code, which can perform a simple *should or shouldn't* render check. The `shouldComponentRender()` method that you've implemented here does exactly that. It compares the new state with the current state with the help of two utility functions:

- `referenceEquality()`: Returns `true` if two arguments are the same reference. This is an extremely fast check to perform.
- `valueEquality()`: Returns `true` if the two array values are the same. This isn't quite as fast because it needs to iterate over the whole array, but it's still faster than the virtual DOM.

The idea for having these two functions separated like this is to handle the fast common case, which is that `setState()` wasn't even called and we have the same array reference, so there's no need to do anything else. If it's not the same object, then we can check for value changes. Even if the values are all the same and it's a new array reference, this method still pays off because it's relatively fast to run and often avoids a trip to the virtual DOM.

Now, let's put this component to work and see what kind of efficiency gains you get:

```
const root =
  ReactDOM.createRoot(document.getElementById("root"));
const myList = React.createRef();

ReactDOM.flushSync(() => {
  root.render(<MyList ref={myList} />);
});

for (let i = 0; i < 100; i++) {
  ReactDOM.flushSync(() => {
    myList.current.setState((state) => ({
      items: [0, ...state.items.slice(1)],
    }));
  });
}
```

You're rendering `<MyList>` over and over, in a loop. Each iteration has 5,000 list items to render. Since the state doesn't change, the call to `shouldComponentUpdate()` returns `false` on every one of these iterations. This is important for performance reasons because there are a lot of them. You're not going to have code that re-renders a component in a tight loop in a real application. This code is meant to stress the rendering capabilities of React. If you were to comment out the `shouldComponentUpdate()` method, you'd see what I mean. Here's what the performance profile looks like for this component:

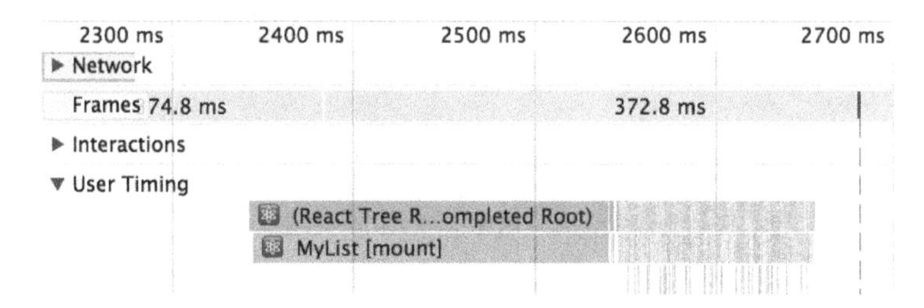

Figure 7.7 – Component performance in React dev tools

The initial render takes the longest (a few hundred milliseconds), but then you have all of these tiny time slices that are completely imperceptible to the user experience. These are the result of `shouldComponentUpdate()` returning `false`. Let's comment out this method now and see how this profile changes:

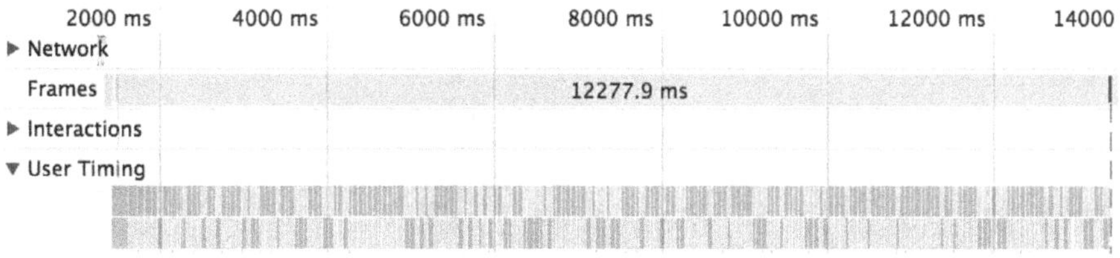

Figure 7.8 – Observing performance changes in React dev tools

Without `shouldComponentUpdate()`, the end result is much larger time slices with a drastically negative impact on user experience. In the next section, we'll try a different approach to optimizing our component rendering in `shouldComponentUpdate()`.

Using metadata to optimize rendering

In this section, you'll learn how to use metadata that's part of the API response to determine whether or not the component should re-render itself. Here's a simple user details component:

```
class MyUser extends React.Component {
  state = {
    modified: new Date(),
    first: "First",
    last: "Last",
  };

  shouldComponentUpdate(props, state) {
    return Number(state.modified) >
      Number(this.state.modified);
  }

  render() {
    const { modified, first, last } = this.state;

    return (
      <section>
        <p>{modified.toLocaleString()}</p>
        <p>{first}</p>
        <p>{last}</p>
      </section>
    );
  }
}
```

The `shouldComponentUpdate()` method is comparing the new, modified state with the old, modified state. This code makes the assumption that the modified value is a date that reflects when the data that was returned by the API was actually modified. The main downside to this approach is that the `shouldComponentUpdate()` method is now tightly coupled with the API data. The advantage is that you get a performance boost in the same way that you would with immutable data.

Here's how this heuristic looks in action:

```
import MyUser from "./MyUser";

const root =
  ReactDOM.createRoot(document.getElementById("root"));
const myUser = React.createRef();

ReactDOM.flushSync(() => {
  root.render(<MyUser ref={myUser} />);
});

myUser.current.setState({
  modified: new Date(),
  first: "First1",
  last: "Last1",
});

setTimeout(() => {
  myUser.current.setState({
    first: "First2",
    last: "Last2",
  });
}, 1000);
```

The `MyUser` component is now entirely dependent on the modified state. If it's not greater than the previous modified value, no render happens.

Here's what the component looks like after it's been rendered twice:

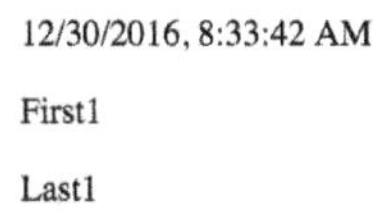

12/30/2016, 8:33:42 AM

First1

Last1

Figure 7.9 – Our component isn't re-rendered because it doesn't need to be

In this section, you learned how to improve the efficiency of your components by using the `shouldComponentUpdate()` life cycle method. Even if your component data hasn't changed, frequently diffing the virtual DOM can cause performance issues. This method exists so that we can build heuristics into our components using an approach that makes sense for our app. In the next section, we'll attempt to render components from other libraries that use an imperative approach.

Rendering imperative components

Everything you've rendered so far in this book has been straightforward declarative HTML. Life is never so simple; sometimes, your React components need to implement some imperative code under the covers.

This is the key: hiding the imperative operations so that the code that renders your component doesn't have to touch it. In this section, you'll implement a simple jQuery UI button React component so that you can see how the relevant life cycle methods help you to encapsulate the imperative code.

Rendering jQuery UI widgets

The jQuery UI widget library implements several widgets on top of standard HTML. It uses a progressive enhancement technique whereby the basic HTML is enhanced in browsers that support newer features. To make these widgets work, you first need to render HTML into the DOM somehow, then, you need to make imperative function calls to create and interact with the widgets.

In this example, you'll create a React button component that acts as a wrapper around the jQuery UI widget. Anyone using the React component shouldn't need to know that, behind the scenes, it's making imperative calls to control the widget. Let's see what the button component looks like:

```
import $ from "jquery";
import "jquery-ui/ui/widgets/button";
import "jquery-ui/themes/base/all.css";

class MyButton extends React.Component {
  componentDidMount() {
    $(this.button).button(this.props);
  }

  componentDidUpdate() {
    $(this.button).button("option", this.props);
  }

  render() {
    return (
      <button
```

```
        onClick={this.props.onClick}
        ref={(button) => {
          this.button = button;
        }}
      />
    );
  }
}
```

The jQuery UI button widget expects an element, so this is what's rendered by the component. An `onClick()` handler from the component props is assigned as well. There's also a `ref` property being used here, which assigns the button argument to `this.button`. The reason this is done is so that the component has direct access to the underlying DOM element of the component. Generally, components don't need access to any DOM elements, but here, you need to issue imperative commands to the element.

For example, in the `componentDidMount()` method, the `button()` function is called and passes its properties from the component. The `componentDidUpdate()` method does something similar and is called when property values change. Now, let's take a look at the button container component:

```
import MyButton from "./MyButton";

class MyButtonContainer extends React.Component {
  componentDidMount() {
    this.setState({
      ...this.props,
      onClick: this.props.onClick.bind(this),
    });
  }

  render() {
    return <MyButton {...this.state} />;
  }
}

MyButtonContainer.defaultProps = {
  onClick: () => {},
};
```

You have a container component that controls the state, which is then passed to `<MyButton>` as properties.

> **Note**
> The `{...data}` syntax is called JSX spread attributes. This allows you to pass objects to elements as attributes. Instead of writing `<User first={data.first} last={data.last} age={data.age} />`, you could shorten it to `<User {...data} />` to get the exact same result.

The component has a default `onClick()` handler function. However, you can pass a different click handler in as a property. Additionally, it's automatically bound to the component context, which is useful if the handler needs to change the button state. Let's look at an example of this:

```
import MyButtonContainer from "./MyButtonContainer";

function onClick() {
  this.setState({ disabled: true });
}

const root =
  ReactDOM.createRoot(document.getElementById("root"));
root.render(
  <section>
    <MyButtonContainer label="Text" />
    <MyButtonContainer
      label="My Button"
      icon="ui-icon-person"
      showLabel={false}
    />
    <MyButtonContainer label="Disable Me" onClick={onClick}
      />
  </section>
);
```

Here, you have three jQuery UI button widgets, each controlled by a React component with no imperative code in sight. Here's how the buttons look:

Figure 7.10 – jQuery UI buttons as React components

In this section, you learned that React components can be used to render imperative components. In order to do so, we need life cycle methods so that we can perform the necessary setup and cleanup operations. In the next section, we'll dig deeper into cleaning up after our components when they're removed.

Cleaning up after components

In this section, you'll learn how to clean up after components. You don't have to explicitly unmount components from the DOM – React handles that for you. There are some things that React doesn't know about and, therefore, cannot clean up for you after the component is removed.

It's for these types of cleanup tasks that the `componentWillUnmount()` life cycle method exists. One use case for cleaning up after React components is asynchronous code.

For example, imagine a component that issues an API call to fetch some data when the component is first mounted. Now, imagine that this component is removed from the DOM before the API response arrives.

Cleaning up asynchronous calls

If your asynchronous code tries to set the state of a component that has been unmounted, nothing will happen. A warning will be logged, and the state won't be set. It's actually very important that this warning is logged; otherwise, you would have a hard time trying to solve subtle race condition bugs.

The correct approach is to create cancellable asynchronous actions. Here's a modified version of the `users()` API function that you implemented in the fetching component data example:

```
import { Promise } from "bluebird";

Promise.config({ cancellation: true });

export function users(fail) {
```

```
  return new Promise((resolve, reject) => {
    setTimeout(() => {
      if (fail) {
        reject(fail);
      } else {
        resolve({
          users: [
            { id: 0, name: "First" },
            { id: 1, name: "Second" },
            { id: 2, name: "Third" }
          ]
        });
      }
    }, 4000);
  });
}
```

Instead of returning a native promise, `users()` returns a promise from the Bluebird library that's been configured to have cancellable behavior. Now, let's take a look at a container component, which has the ability to cancel asynchronous behavior:

```
import root from "./root";
import { users } from "./api";
import UserList from "./UserList";

const onClickCancel = (e) => {
  e.preventDefault();
  root.render(<p>Cancelled</p>,
    document.getElementById("root"));
};

class UserListContainer extends React.Component {
  state = {
    error: null,
    loading: "loading...",
    users: [],
  };
```

```
  componentDidMount() {
    this.job = users();

    this.job.then(
      (result) => {
        this.setState({
          loading: null,
          error: null,
          users: result.users,
        });
      },

      (error) => {
        this.setState({ loading: null, error: error.message
          });
      }
    );
  }

  componentWillUnmount() {
    this.job.cancel();
  }

  render() {
    return <UserList onClickCancel={onClickCancel}
      {...this.state} />;
  }
}
```

The `onClickCancel()` handler actually replaces the user list. This calls the `componentWillUnmount()` method, where you can cancel `this.job`. It's also worth noting that when the API call is made in `componentDidMount()`, a reference to the promise is stored in the component. This is necessary, otherwise, you would have no way to cancel the async call.

Here's what the component looks like when rendered during a pending API call:

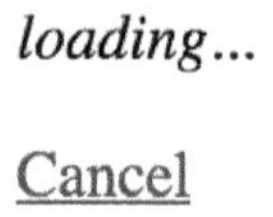

loading...

Cancel

Figure 7.11 – The UI while the API call is pending

Clicking the **Cancel** button causes the `onClickCancel()` function to run, which completely removes `UserListContainer` from the DOM. This, in turn, causes the `componentWillUnmount()` method to run, which will make sure that any pending promises are canceled. Now, we can feel confident that our components can be safely removed, even when they have pending API requests. In the next section, we'll look at life cycle methods to help us control errors in our components.

Containing errors with error boundaries

Error boundaries allow you to handle unexpected component failures. Rather than have every component of your application know how to deal with any errors that it might encounter, error boundaries are a mechanism that you can use to wrap components with error-handling behavior. The best way to think of error boundaries is as `try/catch` syntax for JSX.

Let's revisit the first example from this chapter, where you fetched component data using an API function. The `users()` function accepts a Boolean argument, which, when `true`, causes the promise to reject. This is something that you'll want to handle, but not necessarily in the component that made the API call. In fact, the `UserListContainer` and `UserList` components are already set up to handle API errors like this. The challenge is that if you have lots of components, this is a lot of error-handling code. Furthermore, the error handling is specific to that one API call – what if something else goes wrong?

Here's the modified source for `UserListContainer` that you can use for this example:

```
import { users } from "./api";
import UserList from "./UserList";

class UserListContainer extends React.Component {
  state = {
    error: null,
    loading: "loading...",
    users: [],
```

```
  };

  componentDidMount() {
    users(false).then(
      (result) => {
        this.setState({
          loading: null,
          error: null,
          users: result.users,
        });
      },
      (error) => {
        this.setState({ loading: null, error });
      }
    );
  }

  render() {
    if (this.state.error !== null) {
      throw new Error(this.state.error);
    }
    return <UserList {...this.state} />;
  }
}
```

This component is mostly the same as it was in the first example. The first difference is the call to `users()`, where it's now passing `true`:

```
componentDidMount() {
  users(true).then(
    ...
```

This call will fail, resulting in the `error` state being set. The second difference is in the `render()` method:

```
if (this.state.error !== null) {
  throw new Error(this.state.error);
}
```

Instead of forwarding the error state to the `UserList` component, it's passing the error back to the component tree by throwing an error instead of attempting to render more components. The key design change here is that this component is now making the assumption that there is some sort of error boundary in place further up in the component tree that will handle these errors accordingly.

> **Note**
>
> You might be wondering why the error is thrown in render instead of being thrown when the promise is rejected in `componentDidMount()`. The problem is that fetching data asynchronously like this means that there's no way for the React internals to actually catch exceptions that are thrown from within async promise handlers. The easiest solution for asynchronous actions that could cause a component to fail is to store the error in the component state but to throw the error before actually rendering anything if it's there.

Now, let's create the error boundary itself:

```
class ErrorBoundary extends React.Component {
  state = {
    error: null,
  };

  componentDidCatch(error) {
    this.setState({ error });
  }

  render() {
    if (this.state.error === null) {
      return this.props.children;
    } else {
      return
        <strong>{this.state.error.toString()}</strong>;
    }
  }
}
```

This is where the `componentDidCatch()` life cycle method is utilized by setting the error state of this component when it catches an error. When it's rendered, an error message is rendered if the error state is set. Otherwise, it renders the child components as usual.

Here's how you can use this `ErrorBoundary` component:

```
import ErrorBoundary from "./ErrorBoundary";
import UserListContainer from "./UserListContainer";

const root =
  ReactDOM.createRoot(document.getElementById("root"));
root.render(
  <ErrorBoundary>
    <UserListContainer />
  </ErrorBoundary>
);
```

Any errors that are thrown by `UserListContainer` or any of its children will be caught and handled by `ErrorBoundary`:

Figure 7.12 – What the error looks like in the dev tools console

Now, you can remove the argument that's passed to `users()` in `UserListContainer` to stop it from failing. In the `UserList` component, let's say that you have an error that tries to call `toUpperCase()` on a number:

```
const LoadingMessage = ({ loading }) =>
  loading ? <em>{loading}</em> : null;

function UserList({ error, loading, users }) {
  return (
    <section>
      <LoadingMessage loading={loading} />
      <ul>
        {users.map((user) => (
          <li key={user.id.toUpperCase()}>{user.name}</li>
        ))}
      </ul>
    </section>
  );
}
```

You'll get a different error thrown, but since it's under the same boundary as the previous error, it'll be handled the same way:

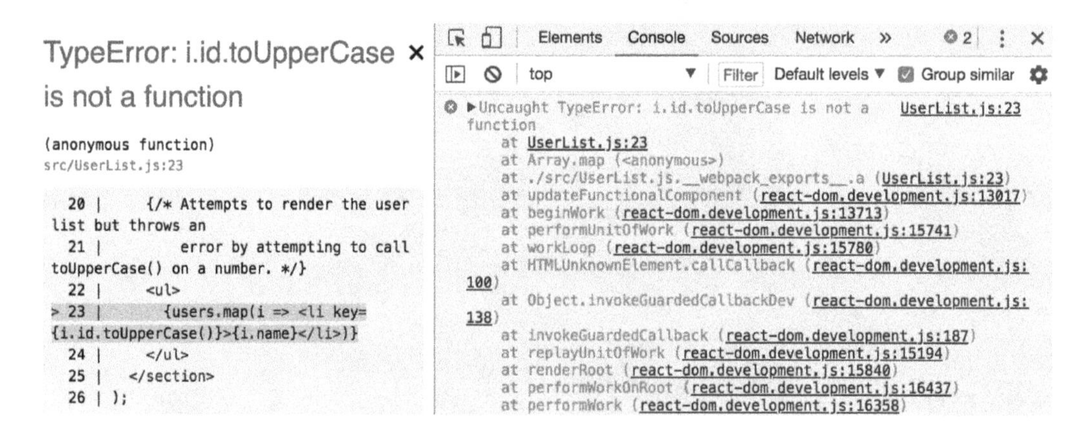

Figure 7.13 – Error being handled by a React error boundary

> **Note**
> If you're running your project with `create-react-app` and `reactscripts`, you might notice an error overlay for every error in your application, even those that are handled by error boundaries. If you close the overlay using the **x** button in the top right, you will be able to see how your component handles the error in your app.

In this section, you learned about the `componentDidCatch()` life cycle method and how it can be used to handle errors in a way that prevents your entire app from crashing. By introducing error boundaries into your application, you have total control over what happens when any piece of your application fails.

Summary

In this chapter, you learned a lot about the life cycle of React components. We started things off with a discussion on why React components need a life cycle in the first place. It turns out that React can't do everything automatically for us, so we need to write some code that's run at the appropriate time during the components' life cycles.

Next, you implemented several components that were able to fetch their initial data and initialize their state from JSX properties. Then, you learned how to implement more efficient React components by providing a `shouldComponentRender()` method.

After that, you learned how to hide the imperative code that some components need to implement and how to clean up after asynchronous behavior. Finally, you learned how to use the new error boundary functionality from React 16.

In the next chapter, you'll learn techniques that help to ensure that your components are being passed the right properties.

Further reading

You can visit the following links for more information:

- React.Component: `https://reactjs.org/docs/react-component.html`
- State and Lifecycle: `https://reactjs.org/docs/state-and-lifecycle.html`

8
Validating Component Properties

In this chapter, you'll learn about property validation in React components. This might seem simple at first glance, but it's an important topic because it leads to bug-free components. We'll start things off with a discussion about predictable outcomes and how this leads to components that are portable throughout the application.

Next, we'll walk through examples of some of the type-checking property validators that come with React. Then, we'll walk through some more complex property-validation scenarios. Finally, we'll wrap this chapter up with an example of how to implement your own custom validators.

The following topics will be covered in this chapter:

- Knowing what to expect
- Promoting portable components
- Simple property validators
- Type and value validators
- Writing custom property validators

Technical requirements

The code files for this chapter can be found on GitHub at `https://github.com/PacktPublishing/React-and-React-Native-4th-Edition/tree/main/Chapter08`.

Knowing what to expect

Property validation in React components is like field validation in HTML forms. The basic premise of validating form fields is letting the user know that they've provided a value that's not acceptable. Ideally, the validation error message is clear enough that the user can easily fix the situation. With React component property validation, you're doing the same thing – making it easy to fix a situation where an unexpected value was provided. Property validation enhances the developer experience, rather than the user experience.

The key aspect of property validation is knowing what's passed into the component as a property value. For example, if you're expecting an array and a Boolean is passed instead, something will probably go wrong. If you validate the property values using the `proptypes` React validation package, then you know that something unexpected was passed. If the component is expecting an array so that it can call the `map()` method, it'll fail if a Boolean value is passed because it has no `map()` method. However, before this failure happens, you'll see the property validation warning.

The idea isn't to fail fast with property validation. It's to provide information to the developer. When property validation fails, you know that something was provided as a component property that shouldn't have been. It's a matter of finding where the value is passed in the code and fixing it.

> **Note**
>
> Fail fast is an architectural property of software in which the system will crash completely rather than continue running in an inconsistent state.

Next, you'll learn how property validation is used to promote portability. These are components that can be used in many places throughout your app.

Promoting portable components

When you know what to expect from your component properties, the context in which the component is used becomes less important. This means that as long as the component is able to validate its property values, it really shouldn't matter where the component is used; it could easily be used by any feature.

If you want a generic component that's portable across application features, you can either write component validation code or you can write defensive code that runs at render time. The challenge with programming defensively is that it dilutes the value of declarative React components. Using React-style property validation, you can avoid writing defensive code. Instead, the property validation mechanism emits a warning when something doesn't pass, informing you that you need to fix something.

> **Defensive Code**
>
> Defensive code is code that needs to account for a number of edge cases during runtime, in a production environment. Coding defensively is necessary when potential problems cannot be detected during development, such as with React component property validation.

Now that you have a better understanding of how property validation assists with writing defensive code and portable components, it's time to implement some property validators.

Simple property validators

In this section, you'll learn how to use the simple property type validators available in the `prop-types` package. Then, you'll learn how to accept any property value as well as to make a property required instead of optional.

Basic type validation

Let's take a look at validators that handle the most primitive types of JavaScript values. You will use these validators frequently, as you'll want to know whether a property is a string or a function, for example. This example will also introduce you to the mechanisms involved in setting up the validation of a component. Here's the component; it just renders some properties using basic markup:

```
import PropTypes from "prop-types";

function MyComponent({
  myString,
  myNumber,
  myBool,
  myFunc,
  myArray,
  myObject,
```

```
}) {
  return (
    <section>
      <p>{myString}</p>
      <p>{myNumber}</p>
      <p>
        <input type="checkbox" defaultChecked={myBool} />
      </p>
      <p>{myFunc()}</p>
      <ul>
        {myArray.map((i) => (
          <li key={i}>{i}</li>
        ))}
      </ul>
      <p>{myObject.myProp}</p>
    </section>
  );
}

MyComponent.propTypes = {
  myString: PropTypes.string,
  myNumber: PropTypes.number,
  myBool: PropTypes.bool,
  myFunc: PropTypes.func,
  myArray: PropTypes.array,
  myObject: PropTypes.object,
};
```

There are two key pieces to the property validation mechanism:

- You have the static `propTypes` property. This is a class-level property, not an instance property. When React finds `propTypes`, it uses this object as the property specification of the component.
- You have the `PropTypes` object from the `prop-types` package, which has several built-in validator functions.

> **Note**
>
> The `PropTypes` object used to be built into React. It was split from React core and moved into the `prop-types` package so that it became opt-in to use – a request by React developers that do not use property validation.

In this example, `MyComponent` has six properties, each with its own type. When you look at the `propTypes` specification, you will see what type of values this component will accept. Let's render this component with some property values:

```
import MyComponent from "./MyComponent";

const root =
  ReactDOM.createRoot(document.getElementById("root"));

const validProps = {
  myString: "My String",
  myNumber: 100,
  myBool: true,
  myFunc: () => "My Return Value",
  myArray: ["One", "Two", "Three"],
  myObject: { myProp: "My Prop" },
};

const invalidProps = {
  myString: 100,
  myNumber: "My String",
  myBool: () => "My Return Value",
  myFunc: true,
  myArray: { myProp: "My Prop" },
  myObject: ["One", "Two", "Three"],
};

function render(props) {
  root.render(<MyComponent {...props} />);
}

render(validProps);
render(invalidProps);
```

The first time <MyComponent> is rendered, it uses the validProps properties. These values all meet the component property specification, so no warnings are logged in the console. The second time around, the invalidProps properties are used, and this fails the property validation because the wrong type is used in every property. The console output should look something like the following:

```
Invalid prop 'myString' of type 'number' supplied to
  'MyComponent',
expected 'string'
Invalid prop 'myNumber' of type 'string' supplied to
  'MyComponent',
expected 'number'
Invalid prop 'myBool' of type 'function' supplied to
  'MyComponent',
expected 'boolean'
Invalid prop 'myFunc' of type 'boolean' supplied to
  'MyComponent', expected
'function'
Invalid prop 'myArray' of type 'object' supplied to
  'MyComponent', expected
'array'
Invalid prop 'myObject' of type 'array' supplied to
  'MyComponent', expected
'object'
TypeError: myFunc is not a function
```

This last error is interesting. We can see that the property validation is complaining about invalid property types. This includes the invalid function that was passed to myFunc. So, despite the type checking that happens on the property, the component will still try to call the value as though it were a function.

Here's what the rendered output looks like:

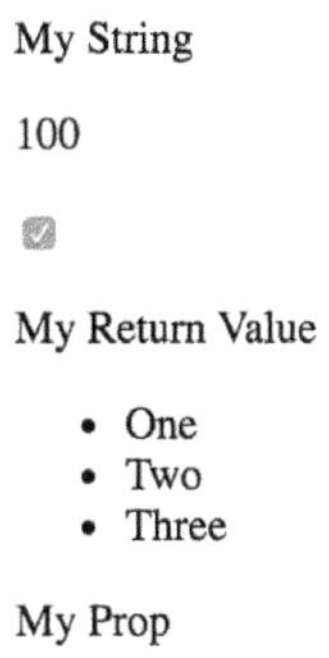

My String

100

☑

My Return Value

- One
- Two
- Three

My Prop

Figure 8.1 – Showing the rendered components that use prop types

> **Note**
>
> Once again, the aim of property validation in React components is to help you discover bugs during development. When React is in production mode, property validation is turned off completely. This means that you don't have to concern yourself with writing expensive property validation code; it'll never run in production. However, the error will still occur, so fix it.

When you validate the type of a given property, nothing is validated if the property isn't passed to the component at all. In the following section, we'll look at how to specify that a property is required and should always be passed.

Requiring values

Let's make some adjustments to the preceding example. The component property specification requires specific types for values, but these are only checked if the property is passed to the component as a JSX attribute. For example, you could have completely omitted the `myFunc` property and it would have been validated. Thankfully, the `PropTypes` functions have a tool that lets you specify that a property must be provided, and it must have a specific type. Here's the modified component:

```
import PropTypes from "prop-types";

function MyComponent({
  myString,
  myNumber,
```

```
  myBool,
  myFunc,
  myArray,
  myObject,
}) {
  return (
    <section>
      <p>{myString}</p>
      <p>{myNumber}</p>
      <p>
        <input type="checkbox" defaultChecked={myBool} />
      </p>
      <p>{myFunc()}</p>
      <ul>
        {myArray.map((i) => (
          <li key={i}>{i}</li>
        ))}
      </ul>
      <p>{myObject.myProp}</p>
    </section>
  );
}

MyComponent.propTypes = {
  myString: PropTypes.string.isRequired,
  myNumber: PropTypes.number.isRequired,
  myBool: PropTypes.bool.isRequired,
  myFunc: PropTypes.func.isRequired,
  myArray: PropTypes.array.isRequired,
  myObject: PropTypes.object.isRequired,
};
```

Not much has changed between this component and the one that you implemented in the preceding section. The main difference is with regard to the specs in `propTypes`. The `isRequired` value is appended to each of the type validators used. So, for instance, `string.isRequired` means that the property value must be a string and that the property cannot be missing. Let's put this component to the test:

```
import MyComponent from "./MyComponent";

const root =
  ReactDOM.createRoot(document.getElementById("root"));

const validProps = {
  myString: "My String",
  myNumber: 100,
  myBool: true,
  myFunc: () => "My Return Value",
  myArray: ["One", "Two", "Three"],
  myObject: { myProp: "My Prop" },
};

const missingProp = {
  myString: "My String",
  myNumber: 100,
  myBool: true,
  myFunc: () => "My Return Value",
  myArray: ["One", "Two", "Three"],
};

function render(props) {
  root.render(<MyComponent {...props} />);
}

render(validProps);
render(missingProp);
```

The first time around, the component is rendered with all of the correct property types. The second time around, the component is rendered without the `myObject` property. The console errors should be as follows:

```
Required prop 'myObject' was not specified in
  'MyComponent'.
Cannot read property 'myProp' of undefined
```

Thanks to the property specification and subsequent error message for `myObject`, it's clear that an object value needs to be provided to the `myObject` property. The last error is because the component assumes that there is an object with `myProp` as a property.

> **Note**
>
> Ideally, you would validate the `myProp` object property in this example since it's directly used in the JSX. The specific properties that are used in the JSX markup for the shape of an object can be validated, as you'll see later in this chapter.

What if you're not exactly sure about the specific type of a given property quite yet? In the next section, we'll look at allowing any value to be passed to a property value while we're still adding a validator for it.

Any property value

The final topic of this section is the any property validator. That is, it doesn't actually care what value it gets – anything is valid, including not passing a value at all. In fact, the `isRequired` validator can be combined with the `any` property validator. For example, if you're working on a component and you just want to make sure that something is passed, but are not sure exactly which type you're going to need yet, you could do something like `myProp: PropTypes.any.isRequired`.

Another reason to have the `any` property validator is for the sake of consistency. Every component should have property specifications. The `any` validator is useful in the beginning when you're not exactly sure what the property type will be. You can at least begin the property spec and then refine it later as things unfold.

Let's take a look at some code:

```
import PropTypes from "prop-types";

function MyComponent({ label, value, max }) {
  return (
```

```
    <section>
      <h5>{label}</h5>
      <progress {...{ max, value }} />
    </section>
  );
}

MyComponent.propTypes = {
  label: PropTypes.any,
  value: PropTypes.any,
  max: PropTypes.any,
};
```

This component doesn't actually validate anything because the three properties in its property spec will accept anything. However, it's a good starting point, because, at a glance, we can see the names of the three properties that this component uses. So, later on, when we decide exactly which types these properties should have, the change is simple. Let's see this component in action:

```
import MyComponent from "./MyComponent";

const root =
  ReactDOM.createRoot(document.getElementById("root"));
root.render(
  <section>
    <MyComponent label="Regular Values" max={20} value={10}
      />
    <MyComponent label="String Values" max="20" value="10"
      />
    <MyComponent
      label={Number.MAX_SAFE_INTEGER}
      max={new Date()}
      value="10"
    />
  </section>
);
```

Strings and numbers are interchangeable in several places. Allowing just one or the other seems overly restrictive. As you'll see in the next section, React has other property validators that allow you to further restrict the property values that are allowed by your component.

Here's what our component looks like when rendered:

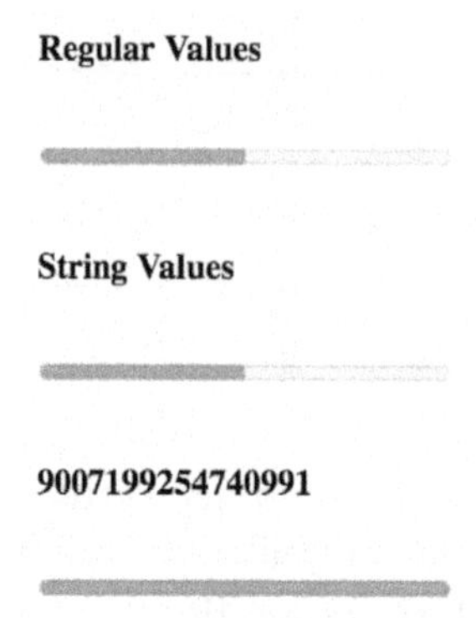

Figure 8.2 – Progress indicators that use different prop type validators

In this section, you learned the basics of property validation for your React components. You can make sure that a property value follows a specific type, that a value is indeed required, and how to allow any value to be passed. In the following section, we'll get into the more specific type and value property validations.

Type and value validators

In this section, you'll learn about the more advanced validator functionality available in the React `prop-types` package. First, you'll learn about the element and node validators that check for values that can be rendered inside HTML markup. Then, you'll see how to check for specific types, beyond the primitive type checking that you just learned about. Finally, you'll implement validation that looks for specific values.

Things that can be rendered

Sometimes, you just want to make sure that a property value is something that can be rendered by JSX markup. For example, if a property value is an array of plain objects, this can't be rendered by putting it in `{}`. You have to map the array items to JSX elements.

This sort of checking is especially useful if your component passes property values to other elements as children. Let's look at an example of what this looks like:

```
import PropTypes from "prop-types";

function MyComponent({ myHeader, myContent }) {
  return (
    <section>
      <header>{myHeader}</header>
      <main>{myContent}</main>
    </section>
  );
}

MyComponent.propTypes = {
  myHeader: PropTypes.element.isRequired,
  myContent: PropTypes.node.isRequired,
};
```

This component has two properties that require values that can be rendered. The `myHeader` property wants an element; this can be any JSX element. The `myContent` property wants a node; this can be any JSX element or any string value. Let's pass this component some values and render it:

```
import MyComponent from "./MyComponent";

const myHeader = <h1>My Header</h1>;
const myContent = <p>My Content</p>;
const root =
  ReactDOM.createRoot(document.getElementById("root"));

root.render(
  <section>
    <MyComponent {...{ myHeader, myContent }} />
    <MyComponent myHeader="My Header" {...{ myContent }} />
    <MyComponent {...{ myHeader }} myContent="My Content"
      />
```

```
      <MyComponent
        {...{ myHeader }}
        myContent={[myContent, myContent, myContent]}
      />
    </section>
  );
```

The `myHeader` property is more restrictive about the values it will accept. The `myContent` property will accept a string, an element, or an array of elements. These two validators are important when passing in child data from properties, as this component does. For example, trying to pass a plain object or a function as a child will not work, and it's best if you check for this situation using a validator.

Here's what this component looks like when rendered:

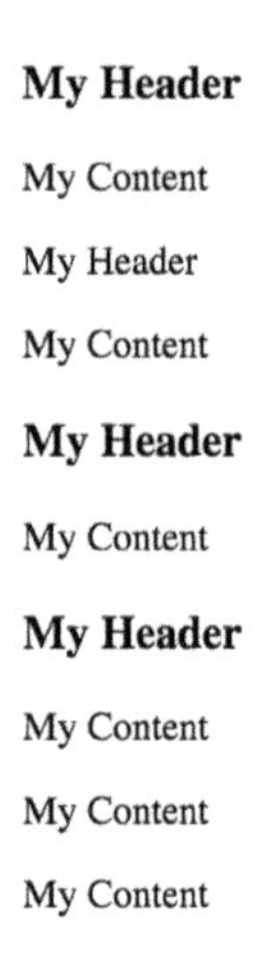

Figure 8.3 – Rendering the header and content components

In the following section, you'll learn how to apply more to your property validators.

Requiring specific types

Sometimes, you need a property validator that checks for a type defined by your application. For example, let's say you have the following user class:

```
const id = (function* () {
  let i = 1;
  while (true) {
    yield i;
```

```
    i += 1;
  }
})();

class MyUser {
  constructor(first, last) {
    this.id = id.next().value;
    this.first = first;
    this.last = last;
  }

  get name() {
    return '${this.first} ${this.last}';
  }
}
```

Now, suppose that you have a component that wants to use an instance of this class as a property value. You would need a validator that checks that the property value is an instance of `MyUser`. Let's implement a component that does just that:

```
import PropTypes from "prop-types";
import MyUser from "./MyUser";

function MyComponent({ myDate, myCount, myUsers }) {
  return (
    <section>
      <p>{myDate.toLocaleString()}</p>
      <p>{myCount}</p>
      <ul>
        {myUsers.map((user) => (
          <li key={user.id}>{user.name}</li>
        ))}
      </ul>
    </section>
  );
}
```

```
MyComponent.propTypes = {
  myDate: PropTypes.instanceOf(Date),
  myCount: PropTypes.oneOfType([PropTypes.string,
    PropTypes.number]),
  myUsers: PropTypes.arrayOf(PropTypes.instanceOf(MyUser)),
};
```

This component has three properties that require specific types, each going beyond the basic type validators that you've seen so far in this chapter. Let's walk through these now:

- `myDate` requires an instance of `Date`. It uses the `instanceOf()` function to build a validator function that ensures the value is a `Date` instance.
- `myCount` requires that the value either be a number or a string. This validator function is created by combining `oneOfType`, `PropTypes.number()`, and `PropTypes.string()`.
- `myUsers` requires an array of `MyUser` instances. This validator is built by combining `arrayOf()` and `instanceOf()`. This example illustrates the number of scenarios that you can handle by combining the property validators provided by React.

Here's what the rendered output looks like:

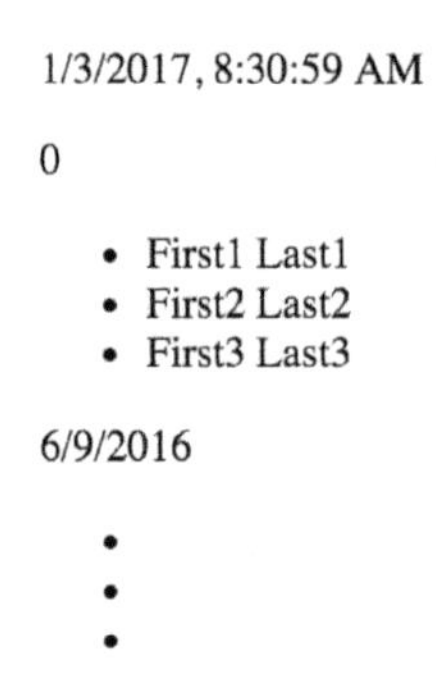

Figure 8.4 – The rendered output of components

In the next section, we'll look at validating the actual values that are passed to component properties.

Requiring specific values

I've focused on validating the type of property values so far, but that's not always what you'll want to check for. Sometimes, specific values matter. Let's see how we can validate specific property values:

```
import PropTypes from "prop-types";

const levels = new Array(10).fill(null).map((v, i) => i +
  1);
const userShape = {
  name: PropTypes.string,
  age: PropTypes.number,
};

function MyComponent({ level, user }) {
  return (
    <section>
      <p>{level}</p>
      <p>{user.name}</p>
      <p>{user.age}</p>
    </section>
  );
}

MyComponent.propTypes = {
  level: PropTypes.oneOf(levels),
  user: PropTypes.shape(userShape),
};
```

The level property is expected to be a number from the `levels` array. This is easy to validate using the `oneOf()` function. The `user` property is expecting a specific shape. A shape is the expected properties and types of an object. The `userShape` defined in this example requires a `name` string and an `age` number. The key difference between `shape()` and `instanceOf()` is that you don't necessarily care about the type. You might only care about the values that are used in the JSX of the component.

Let's take a look at how this component is used:

```
import MyComponent from "./MyComponent";

const root =
  ReactDOM.createRoot(document.getElementById("root"));
root.render(
  <section>
    <MyComponent level={10} user={{ name: "Name", age: 32
      }} />
    <MyComponent user={{ name: "Name", age: 32, online:
      false }} />
    <MyComponent level={11} user={{ name: "Name", age: "32"
      }} />
  </section>
);
```

Here's what the component looks like when it's rendered:

10

Name

32

Name

32

11

Name

32

Figure 8.5 – The rendered component output

In this section, you learned about the property validation tools that are available to validate very precise requirements regarding the property types and property values. In the following section, you'll learn how to build your own property validators.

Writing custom property validators

In this final section, you'll learn how to build your own custom property validation functions and apply them to the property specification. Generally speaking, you should only implement your own property validator if you absolutely have to. The default validators available in `prop-types` cover a wide range of scenarios.

However, sometimes, you need to make sure that very specific property values are passed to the component. Remember, these will not be run in production mode, so it's perfectly acceptable for a validator function to iterate over collections. Let's implement some custom validator functions:

```
function MyComponent({ myArray, myNumber }) {
  return (
    <section>
      <ul>
        {myArray.map((i) => (
          <li key={i}>{i}</li>
        ))}
      </ul>
      <p>{myNumber}</p>
    </section>
  );
}

MyComponent.propTypes = {
  myArray: (props, name, component) =>
    Array.isArray(props[name]) && props[name].length
      ? null
      : new Error('${component}.${name}: expecting non-
        empty array'),

  myNumber: (props, name, component) =>
    Number.isFinite(props[name]) &&
    props[name] > 0 &&
    props[name] < 100
      ? null
      : new Error(
```

```
            `${component}.${name}: expecting number between 1
              and 99`
        ),
};
```

The `myArray` property expects a non-empty array, and the `myNumber` property expects a number that's greater than 0 and less than 100. Let's try passing these validators some data:

```
import MyComponent from "./MyComponent";

const root =
  ReactDOM.createRoot(document.getElementById("root"));
root.render(
  <section>
    <MyComponent
      myArray={["first", "second", "third"]}
      myNumber={99}
    />
    <MyComponent myArray={[]} myNumber={100} />
  </section>
);
```

The first element renders just fine, as both of the validators return null. However, the empty array and the number 100 cause both validators to return errors, like so:

```
MyComponent.myArray: expecting non-empty array
MyComponent.myNumber: expecting number between 1 and 99
```

Here's what the rendered output looks like:

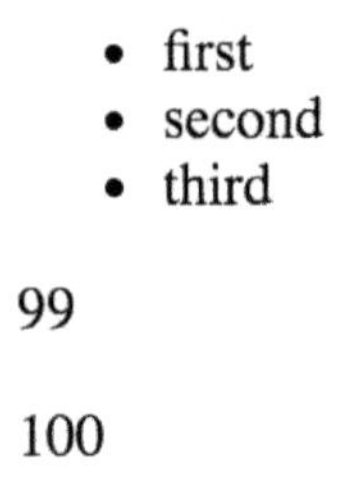

- first
- second
- third

99

100

Figure 8.6 – Rendering components with prop types

In this section, you learned how to construct your own functions that are given a number of arguments so that you can validate the property value. As long as the function returns `true` when the property value is considered valid, you can do almost any kind of validation that you can imagine. These functions can then be passed to the `propTypes` object, just like any of the built-in property validators.

Summary

The focus of this chapter has been React component property validation. When you implement property validation, you know what to expect; this promotes portability. The component doesn't care how the property values are passed to it, just as long as they're valid.

Then, you worked on several examples that used the basic React validators to check primitive JavaScript types. You also learned that if a property is required, it must be made explicit. Next, you learned how to validate more complex property values by combining the built-in validators that come with React.

Finally, you implemented your own custom validator functions to perform validation that goes beyond what's possible with the `prop-types` validators. In the next chapter, you'll learn how to handle navigation using React routes.

Further reading

To find out more about type checking with `PropTypes`, you can refer to `https://reactjs.org/docs/typechecking-with-proptypes.html`.

9
Handling Navigation with Routes

Almost every web application requires routing: the process of responding to a URL, based on a set of route handler declarations. In other words, this is a mapping from the URL to rendered content. However, this task is more involved than it seems at first. This is why you're going to leverage the `react-router` package in this chapter, the de facto routing tool for React.

First, you'll learn the basics of declaring routes using JSX syntax. Then, you'll learn about the dynamic aspects of routing, such as dynamic path segments and query parameters. Next, you'll implement links using components from `react-router`.

Here are the high-level topics that we'll cover in this chapter:

- Declaring routes
- Handling route parameters
- Using link components

Technical requirements

You can find the code files for this chapter on GitHub at `https://github.com/PacktPublishing/React-and-React-Native-4th-Edition/tree/main/Chapter09`.

Declaring routes

With `react-router`, you can collocate routes with the content that they render. In this section, you'll see that this is done using JSX syntax that defines routes.

You'll create a basic hello world example route so that you can see what routes look like in React applications. Then, you'll learn how you can organize your route declarations by feature instead of in a monolithic module. Finally, you'll implement a common parent-child routing pattern.

Hello route

Let's create a simple route that renders a simple component. First, we have a small React component that we want to render when the route is activated:

```
function MyComponent() {
  return <p>Hello Route!</p>;
}
```

Next, let's look at the route definition:

```
import * as React from "react";
import * as ReactDOM from "react-dom";
import { BrowserRouter as Router, Route, Routes } from
  "react-router-dom";
import MyComponent from "./MyComponent";

const root =
  ReactDOM.createRoot(document.getElementById("root"));

root.render(
  <Router>
    <Routes>
      <Route path="/" element={<MyComponent />} />
```

```
    </Routes>
  </Router>
);
```

The `Router` component is the top-level component of the application. Let's break it down to find out what's happening within the router.

You have the actual routes declared as `<Route>` elements. There are two key properties of any route: `path` and `element`. When the path is matched against the active URL, the component is rendered. But where is it rendered, exactly? The `Router` component doesn't actually render anything itself; it's responsible for managing how other components are rendered based on the current URL. Sure enough, when you look at this example in a browser, `<MyComponent>` is rendered as expected:

Hello Route!

Figure 9.1 – The rendered output of our component

When the `path` property matches the current URL, `<Route>` is replaced by the `element` property value. In this example, the route is replaced with `<MyComponent>`. If a given route doesn't match, nothing is rendered.

Decoupling route declarations

The difficulty with routing happens when your application has dozens of routes declared within a single module since it's more difficult to mentally map routes to features.

To help with this, each top-level feature of the application can define its own routes. This way, it's clear which routes belong to which feature. So, let's start with the `App` component:

```
import React from "react";
import { BrowserRouter as Router, Route, Routes } from
  "react-router-dom";
import Layout from "./Layout";
import oneRoutes from "./one";
import Redirect from "./Redirect";
import twoRoutes from "./two";

export default () => (
  <Router>
    <Routes>
```

```
        <Route path="/" element={<Layout />}>
          <Route index element={<Redirect path="/one" />} />
          {oneRoutes}
          {twoRoutes}
        </Route>
      </Routes>
    </Router>
);
```

In this example, the application has two features: `one` and `two`. These are imported as components and rendered inside `<Router>`. The first child in this router is actually a redirect. This means that when the app first loads the URL, `/`, the `<Redirect>` component will send the user to `/one`. The render property is an alternative to the component property when you need to call a function to render content. You're using it here because you need to pass the property to `<Redirect>`.

This module will only get as big as the number of application features, instead of the number of routes, which could be substantially larger. Let's take a look at one of the feature routes:

```
import * as React from "react";
import { Route, Outlet } from "react-router-dom";
import Redirect from "../Redirect";
import First from "./First";
import Second from "./Second";

const routes = (
  <Route path="/one" element={<Outlet />}>
    <Route index element={<Redirect path="/one/1" />} />
    <Route path="/one/1" element={<First />} />
    <Route path="/one/2" element={<Second />} />
  </Route>
);

export default routes;
```

This module, `one/index.js`, exports an element with three routes:

- When the `/one` path is matched, redirect to `/one/1`.
- When the `/one/1` path is matched, render the `First` component.
- When the `/one/2` path is matched, render the `Second` component.

This follows the same pattern as the `App` component for the / path. Often, your application doesn't actually have content to render at the root of a feature, or at the root of the application itself. This pattern allows you to send the user to the appropriate route and the appropriate content. Here's what you'll see when you first load the app:

Feature 1, page 1

Figure 9.2 – The contents of page 1

The second feature follows the exact same pattern as the first. Here's what the `First` component looks like:

```
export default function First() {
  return <p>Feature 1, page 1</p>;
}
```

Each feature, in this example, uses the same minimal rendered content. These components are ultimately what the user needs to see when they navigate to a given route. By organizing routes this way, you've made your features self-contained with regard to routing.

In the following section, you'll learn how to further organize your routes into parent-child relationships.

Handling route parameters

The URLs that you've seen so far in this chapter have all been static. Most applications will use both static and dynamic routes. In this section, you'll learn how to pass dynamic URL segments to your components, how to make these segments optional, and how to get query string parameters.

Resource IDs in routes

One common use case is to make the ID of a resource part of the URL. This makes it easy for your code to get the ID and then make an API call that fetches the relevant resource data. Let's implement a route that renders a user detail page. This will require a route that includes the user ID, which then needs to somehow be passed to the component so that it can fetch the user.

Let's start with the `App` component that declares the routes:

```
import { BrowserRouter as Router, Routes, Route } from
  "react-router-dom";
import UsersContainer from "./UsersContainer";
import UserContainer from "./UserContainer";

function App() {
  return (
    <Router>
      <Routes>
        <Route path="/" element={<UsersContainer />} />
        <Route path="/users/:id" element={<UserContainer
          />} />
      </Routes>
    </Router>
  );
}
```

The `:` syntax marks the beginning of a URL variable. The `id` variable will be passed to the `UserContainer` component – here's how it's implemented:

```
import React, { useState, useEffect } from "react";
import { useParams } from "react-router-dom";
import User from "./User";
import { fetchUser } from "./api";

function UserContainer() {
  const params = useParams();
  const [error, setError] = useState();
  const [first, setFirst] = useState();
```

```
  const [last, setLast] = useState();
  const [age, setAge] = useState();

  useEffect(() => {
    fetchUser(Number(params.id)).then(
      (user) => {
        setError(null);
        setFirst(user.first);
        setLast(user.last);
        setAge(user.age);
      },
      (error) => {
        setError(error);
        setFirst(null);
        setLast(null);
        setAge(null);
      }
    );
  }, [params.id]);

  return <User {...{ error, first, last, age }} />;
}

export default UserContainer;
```

The `useParams()` hook is used to get any dynamic parts of the URL. In this case, you're interested in the `id` parameter. Then, you pass the number version of this value to the `fetchUser()` API call. If the URL is missing the segment completely, then this code won't run at all; the router will revert back to the / route. However, no type checking is done at the route level, which means it's up to you to handle non-numbers being passed where numbers are expected, and so on.

In this example, the type cast operation will result in a 500 error if the user navigates to, for example, `/users/one`. You could write a function that type-checks parameters and, instead of failing with an exception, responds with a 404: Not found error. In any case, it's up to the application, and not the `react-router` library, to provide a meaningful failure mode.

Now, let's take a look at the API functions that were used in this example:

```
const users = [
  { first: "First 1", last: "Last 1", age: 1 },
  { first: "First 2", last: "Last 2", age: 2 }
];

export function fetchUsers() {
  return new Promise((resolve, reject) => {
    resolve(users);
  });
}

export function fetchUser(id) {
  return new Promise((resolve, reject) => {
    const user = users[id];

    if (user === undefined) {
      reject('User ${id} not found');
    } else {
      resolve(user);
    }
  });
}
```

The `fetchUsers()` function is used by the `UsersContainer` component to populate the list of user links. The `fetchUser()` function will find and resolve a value from the `users` array of the mock data if the promise is rejected. If rejected, the error handling behavior of the `UserContainer` component is invoked.

Here is the `User` component, which is responsible for rendering the user details:

```
import React from "react";
import PropTypes from "prop-types";

const Error = ({ error }) =>
  error ? (
    <p>
```

```
      <strong>{error}</strong>
    </p>
  ) : null;
const Text = ({ children }) => (children ?
  <p>{children}</p> : null);

function User({ error, first, last, age }) {
  return (
    <section>
      <Error error={error} />
      <Text>{first}</Text>
      <Text>{last}</Text>
      <Text>{age}</Text>
    </section>
  );
}

User.propTypes = {
  error: PropTypes.string,
  first: PropTypes.string,
  last: PropTypes.string,
  age: PropTypes.number,
};

export default User;
```

When you run this app and navigate to `/`, you should see a list of users that looks like this:

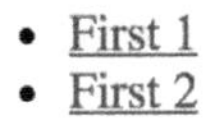

- First 1
- First 2

Figure 9.3 – The contents of the app home page

Clicking on the first link should take you to `/users/0`, which looks like this:

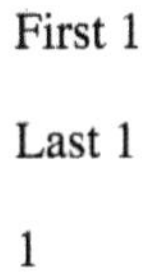

First 1

Last 1

1

Figure 9.4 – The contents of the user page

If you navigate to a user that doesn't exist, for example, `/users/2`, here's what you'll see:

User 2 not found

Figure 9.5 – When a user isn't found

The reason that you get this error message instead of a 500 error is that the API endpoint knows how to deal with missing resources:

```
if (user === undefined) {
  reject('User ${id} not found');
}
```

This results in `UserContainer` setting its error state:

```
fetchUser(Number(params.id)).then(
  (user) => {
    setError(null);
    setFirst(user.first);
    setLast(user.last);
    setAge(user.age);
  },
  (error) => {
    setError(error);
    setFirst(null);
    setLast(null);
    setAge(null);
  }
);
```

This then results in the `User` component rendering the error message:

```
const Error = ({ error }) =>
  error ? (
    <p>
      <strong>{error}</strong>
    </p>
  ) : null;
const Text = ({ children }) => (children ?
```

```
    <p>{children}</p> : null);

function User({ error, first, last, age }) {
  return (
    <section>
      <Error error={error} />
      <Text>{first}</Text>
      <Text>{last}</Text>
      <Text>{age}</Text>
    </section>
  );
}
```

Since the `error` property value is a string, the `Error` component will render the error message. In the next section, we'll look at defining optional route parameters.

Optional parameters

Sometimes, we need optional URL path values and query parameters. URLs work best for simple options, and query parameters work best if there are many values that the component can use.

Let's implement a user list component that renders a list of users. Optionally, you want to be able to sort the list in descending order. Let's make this an optional path segment in the route definition for this page:

```
import * as React from "react";
import * as ReactDOM from "react-dom";
import { BrowserRouter as Router, Route, Routes } from
  "react-router-dom";
import UsersContainer from "./UsersContainer";

const root =
  ReactDOM.createRoot(document.getElementById("root"));
root.render(
  <Router>
    <Routes>
      <Route path="/users">
        <Route path=":desc" element={<UsersContainer />} />
```

```
        <Route path="" element={<UsersContainer />} />
      </Route>
    </Routes>
  </Router>
);
```

The `:` syntax marks a variable, and in this case, we've called it `desc`. To make this variable optional, we've added another route that doesn't have this variable but renders the same component.

It's also up to the component to handle any query strings provided to it. So, while the route declaration doesn't provide a mechanism to define accepted query strings, the router will still pass the raw query string to the component. Let's take a look at the user list container component:

```
import React, { useState, useEffect } from "react";
import { useParams, useSearchParams } from "react-router-
  dom";
import Users from "./Users";
import { fetchUsers } from "./api";

function UsersContainer() {
  const [users, setUsers] = useState([]);
  const params = useParams();
  const [search] = useSearchParams();

  useEffect(() => {
    const desc = params.desc === "desc" ||
      !!search.get("desc");

    fetchUsers(desc).then((users) => {
      setUsers(users);
    });
  }, [params, search]);

  return <Users users={users} />;
```

```
}

export default UsersContainer;
```

This component looks for either `params.desc` or `search.desc`. It uses this as an argument to the `fetchUsers()` API to determine the sort order.

Here's what the `Users` component looks like:

```
import * as React from "react";
import PropTypes from "prop-types";

function Users({ users }) {
  return (
    <ul>
      {users.map((user) => (
        <li key={user}>{user}</li>
      ))}
    </ul>
  );
}

Users.propTypes = {
  users: PropTypes.array.isRequired,
};

export default Users;
```

Here is what's rendered when you navigate to `/users`:

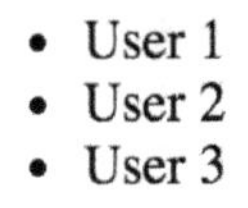

- User 1
- User 2
- User 3

Figure 9.6 – Rendering the user list in default order

If you include the descending parameter by navigating to `/users/desc`, here's what you get:

- User 3
- User 2
- User 1

Figure 9.7 – Rendering the user list in descending order

In this section, you learned about parameters in routes. Perhaps the most common pattern is to have the ID of a resource in your app as part of the URL, which means that components need to be able to parse out this information in order to interact with the API. You also learned about optional parameters in routes – these aren't always required because the component will use default values when they're not provided. In the next section, you'll learn about link components.

Using link components

In this section, you'll learn how to create links. You might be tempted to use the standard `<a>` elements to link to pages controlled by `react-router`. The problem with this approach is that these links will try to locate the page on the backend by sending a `GET` request. This isn't what you want because the route configuration is already in the browser.

First, you'll see an example that illustrates how `<Link>` elements are just like `<a>` elements in most ways. Then, you'll see how to build links that use URL parameters and query parameters.

Basic linking

The idea of links in React apps is that they point to routes that point to components, which render new content. The `Link` component also takes care of the browser history API and looks up route-component mappings. Here's an application component that renders two links:

```
import * as React from "react";
import {
  BrowserRouter as Router,
  Route,
  Routes,
  Link,
  Outlet
} from "react-router-dom";
```

```
import Layout from "./Layout";
import First from "./First";
import Second from "./Second";

function Layout() {
  return (
    <>
      <nav>
        <p>
          <Link to="first">First</Link>
        </p>
        <p>
          <Link to="second">Second</Link>
        </p>
      </nav>
      <main>
        <Outlet />
      </main>
    </>
  );
}

function App() {
  return (
    <Router>
      <Routes>
        <Route path="/" element={<Layout />}>
          <Route path="/first" element={<First />} />
          <Route path="/second" element={<Second />} />
        </Route>
      </Routes>
    </Router>
  );
}
```

The `to` property specifies the route to activate when clicked. In this case, the application has two routes – `/first` and `/second`. Here is what the rendered links look like:

First

Second

Figure 9.8 – Links to the first and second pages of the app

When you click the first link, the page content changes to look like this:

First

Figure 9.9 – The first page when the app is rendered

Now that you can use `Link` components to render links to basic paths, it's time to learn about building dynamic links with parameters.

URL and query parameters

Constructing the dynamic segments of a path that is passed to `<Link>` involves string manipulation. Everything that's part of the path goes to the `to` property. This means that you have to write more code to construct the string, but it also means less behind-the-scenes magic happening in the router.

Let's create a simple component that will echo back whatever is passed to the echo URL segment or the echo query parameter:

```
import React from "react";
import { useParams, useSearchParams } from "react-router-
  dom";

function Echo() {
  const params = useParams();
  const [searchParams] = useSearchParams();

  return <h1>{params.msg || searchParams.get("msg")}</h1>;
}

export default Echo;
```

In order to get search parameters that were passed to a route, you can use the `useSearchParams()` hook, which gives you a `URLSearchParams` object. In this case, we can call `searchParams.get("msg")` to get the parameter we need.

Now, let's take a look at the `App` component that renders two links. The first will build a string that uses a dynamic value as a URL parameter. The second will use `URLSearchParams` to build the query string portion of the URL:

```
import React from "react";
import PropTypes from "prop-types";
import { Link } from "react-router-dom";

export default function App({ children }) {
  return <section>{children}</section>;
}

App.propTypes = {
  children: PropTypes.node.isRequired
};

const param = "From Param";
const query = new URLSearchParams({ msg: "From Query" });

App.defaultProps = {
  children: (
    <section>
      <p>
        <Link to={'echo/${param}'}>Echo param</Link>
      </p>
      <p>
        <Link to={'echo?${query.toString()}'}
          query={query}>
          Echo query
        </Link>
      </p>
    </section>
  )
};
```

Here's what the two links look like when they're rendered:

Echo param

Echo query

Figure 9.10 – Different types of link parameters

The param link takes you to `/echo/From Param`, which looks like this:

From Param

Figure 9.11 – The param version of the page

The query link takes you to `/echo?echo=From+Query`, which looks like this:

From Query

Figure 9.12 – The query version of the page

In this section, you learned about using the `Link` component to render links in your application. You also learned how to build dynamic links that pass parameters to URLs.

Summary

In this chapter, you learned about routing in React applications. The job of a router is to render content that corresponds to a URL. The `react-router` package is the standard tool for this job. You learned how routes are JSX elements, just like the components they render. Sometimes, you need to split routes into feature-based modules. A common pattern for structuring page content is to have a parent component that renders the dynamic parts as the URL changes.

Then, you learned how to handle the dynamic parts of URL segments and query strings. You also learned how to build links throughout your application using the `<Link>` element.

In the next chapter, you'll learn how to split your code into smaller chunks using lazy components.

Further reading

Refer to the following links for more information:

- React Router: `https://reactrouter.com/`
- `URLSearchParams`: `https://developer.mozilla.org/en-US/docs/Web/API/URLSearchParams`

10
Code Splitting Using Lazy Components and Suspense

Code splitting has been happening in React applications for some time now, long before there was any official support in the React API. With the latest version of React, there are new APIs that we can use that directly support code-splitting scenarios. Code splitting is necessary when you have larger applications with a lot of JavaScript code that must be delivered to a browser.

Big monolithic JavaScript bundles that house an entire application can create usability issues on initial page loads due to longer load times. With code splitting, we have more fine-grained control over how code makes its way from the server to the browser. This means more opportunities for us to properly handle load-time **User Experience** (**UX**).

In this chapter, you'll learn how to do this in your React applications by using the `lazy()` API and the **Suspense** components, two recent additions to React. Once you understand how these two pieces work, you'll be able to completely integrate code splitting into your applications.

We'll cover the following topics in this chapter:

- Using the `lazy()` API
- Using the `Suspense` component
- Avoiding lazy components
- Exploring lazy pages and routes

Technical requirements

You can find the code files of this chapter on GitHub at `https://github.com/PacktPublishing/React-and-React-Native-4th-Edition/tree/main/Chapter10`.

Using the lazy API

There are two pieces involved with using the new `lazy()` API in React. First, there's bundling components into their own separate files so that they can be downloaded by the browser separately from other parts of the application. Second, once you have created the bundles, you can build React components that are lazy – they don't download anything until the first time they're rendered. Let's look at both of these.

Dynamic imports and bundles

The code examples in this book use the `create-react-app` tooling for creating bundles. The nice thing about this approach is that you don't have to maintain any bundle configuration. Instead, bundles are created for you automatically, based on how you import your modules. If you're using the `import` statement everywhere, your app will be downloaded all at once in one bundle. When your app gets bigger, there will likely be features that some users never use or don't use as frequently as others. You can use the `import()` function to import modules on demand. By using this function, you're telling webpack to create a separate bundle for the code that you're importing dynamically.

Let's look at a simple component that we might want to bundle separately from the rest of the application:

```
export default function MyComponent() {
  return <p>My Component</p>;
}
```

Now, let's take a look at how we would import this module dynamically using the `import()` function, resulting in a separate bundle:

```
function App() {
  const [MyComponent, setMyComponent] = React.useState
    (() => () => null);

  React.useEffect(() => {
    import("./MyComponent").then((module) => {
      setMyComponent(() => module.default);
    });
  }, []);

  return <MyComponent />;
}
```

When you run this example, you'll see the `<p>` text rendered right away. If you open the browser dev tools and look at the network requests, you'll notice that a separate call is made to fetch the bundle containing the `MyComponent` code. This happens because of the call to `import("./MyComponent")`. The `import()` function returns a promise that resolves with the module object. Since we need the default export to access `MyComponent`, we reference `module.default` when we call `setMyComponent()`.

The reason why we're setting a component as the `MyComponent` state is that when the `App` component renders for the first time, we don't have the `MyComponent` code loaded yet. Once it loads, `MyComponent` will reference the proper value, which results in the correct text being rendered.

Now that you have an idea of how bundles get created and are fetched by the app, it's time to see how the `lazy()` API greatly simplifies this process for us.

Making components lazy

Instead of manually handling the promise returned by `import()` by returning the default export and setting state, you can lean on the `lazy()` API. This function takes a function that returns an `import()` promise. The return value is a lazy component that you can just render. Let's modify the `App` component to use this API:

```
import MyPage from "./MyPage";

function App() {
```

```
  return (
    <Suspense fallback={"loading..."}>
      <MyPage />
    </Suspense>
  );
}
```

The `MyComponent` value is created by calling `lazy()`, passing in the dynamic module import as an argument. Now, you have a separate bundle for your component and a lazy component that loads the bundle when it's first rendered.

In this section, you learned how code splitting works. You learned that the `import()` function handles bundle creation for you. You also learned that the `lazy()` API makes your components lazy and handles all of the gritty work of importing components for you. But there's one last thing we need, the `Suspense` component, to help display placeholders while components are loading.

Using the Suspense component

In this section, we'll explore some of the more common usage scenarios of the `Suspense` component. We'll look at where to place `Suspense` components in your component tree, how to simulate latency when fetching bundles, and some of the options available to us to use as the fallback content.

Top-level Suspense components

Lazy components need to be rendered inside of a `Suspense` component. However, they do not have to be direct children of `Suspense` though, which is important because this means that you can have one `Suspense` component handle every lazy component in your app. Let's illustrate this concept with an example. Here's a component that we would like to bundle separately and use lazily:

```
export default function MyFeature() {
  return <p>My Feature</p>;
}
```

Next, let's make the `MyFeature` component lazy and render it inside of a `MyPage` component:

```
const MyFeature = React.lazy(() => import("./MyFeature"));

function MyPage() {
  return (
    <>
      <h1>My Page</h1>
      <MyFeature />
    </>
  );
}
```

Here, we're using the `lazy()` API to make the `MyFeature` component lazy. This means that when the `MyPage` component is rendered, the code bundle that contains `MyFeature` will be downloaded because `MyFeature` was also rendered. What's important to note with the `MyPage` component is that it is rendering a lazy component (`MyFeature`) but isn't rendering a `Suspense` component. This is because our hypothetical app has many page components, each with its own lazy components. Having each of these components render its own `Suspense` component would be redundant. Instead, we can render one `Suspense` component inside of our `App` component, like so:

```
import MyPage from "./MyPage";

function App() {
  return (
    <React.Suspense fallback={"loading..."}>
      <MyPage />
    </React.Suspense>
  );
}
```

While the `MyFeature` code bundle is being downloaded, `<MyPage>` is replaced with the fallback text passed to `Suspense`. So, even though `MyPage` isn't lazy itself, it renders a lazy component that `Suspense` knows about and will replace its children with the fallback content while this happens.

So far, we haven't really been able to see the fallback content that displays while our lazy components load their code bundles. This is because when developing locally, these bundles load almost instantly. In the next section, we'll look at an approach to simulate latency when loading code bundles.

Simulating latency

The whole idea with the `lazy()` and `Suspense` APIs is to provide a better user experience for both of the following:

- The initial load, by splitting code into bundles so that the whole app doesn't have to be downloaded upfront
- Providing consistent fallback content while code bundles load

Unless we can experience latency similar to what our users are likely to experience, we have no idea how effective our use of these APIs is. One way to address this issue is to simulate latency in the same way that you might simulate latency in a mock API call. In the mock function that returns a promise, you use a `setTimeout()` call that resolves the promise after some time – say, 3 seconds. Because the `import()` function returns a promise, we can use this to our advantage.

Here's an updated version of the `MyPage` component from the top-level suspense component example:

```
const MyFeature = React.lazy(() =>
  Promise.all([
    import("./MyFeature"),
    new Promise((resolve) => {
      setTimeout(() => {
        resolve();
      }, 3000);
    }),
  ]).then(([m]) => m)
);

function MyPage() {
  return (
    <>
      <h1>My Page</h1>
```

```
            <MyFeature />
        </>
    );
}
```

Now when you load the example, you'll actually get to see the loading text for about 3 seconds before it's replaced with the `MyPage` content. Instead of just returning the promise from `import()`, we're building a new promise using `Promise.all()`. This method returns a promise that resolves when all of the promises that are passed to it have been resolved. In this example, we're passing two promises to `Promise.all()`. The first is the promise returned by `import()`, which eventually resolves the module object from the code bundle once it's downloaded. The problem is that this resolves immediately when doing local development. The second promise that's passed to `Promise.all()` is how we simulate latency by not resolving the promise for 3 seconds.

The last thing we need to do is make sure that it's the module that's resolved since this is what `lazy()` is expecting. When `Promise.all()` resolves, all of the resolved values are passed as an array to `.then()`. To address this, we add our own `.then()` that returns the first array argument, which is the module that `lazy()` needs.

Now that we can actually see our loading fallback content in action, let's work on making this content a little bit more visually appealing.

Working with spinner fallbacks

The simplest fallback that you can use with the `Suspense` component is some text that indicates to the user that something is happening. The `fallback` property can be any valid React element, which means that we can enhance the fallback to be something more visually appealing. For example, the `react-spinners` package has a selection of spinner components, all of which can be used as a fallback with `Suspense`.

Let's modify the `App` component from the *Simulating latency* section to include a spinner from the `react-spinners` package as the `Suspense` fallback:

```
import { FadeLoader } from "react-spinners";
import MyPage from "./MyPage";

function App() {
  return (
    <React.Suspense fallback={<FadeLoader
      color={"lightblue"} size={150} />}>
```

```
      <MyPage />
    </React.Suspense>
  );
}
```

The `FadeLoader` component will render a spinner that we've configured with a `color` value of `lightblue` and a size of `150` pixels. The rendered element of the `FadeLoader` component is passed to the fallback property. Since we're simulating latency, you should be able to see the spinner when you first load the app:

Figure 10.1 – The image rendered by the loader component

Now, instead of text, we're showing an animated spinner. This likely provides a user experience that your users are more accustomed to. The `react-spinners` package has several spinners for you to choose from, each of which has several configuration options. There are other spinner libraries that you can use or implement on your own.

In this section, you learned that you can use a single `Suspense` component that will display its fallback content for any lazy components that are lower in the tree. You learned how to simulate latency during local development so that you can experience what your users will experience with your `Suspense` fallback content. Finally, you learned how to use components from other libraries as the fallback content to provide something that looks better than plain text.

In the next section, you'll learn about why it doesn't make sense to make every component in your app a lazy component.

Avoiding lazy components

It might be tempting to make most of your React components lazy components that live in their own bundle. After all, there isn't much extra work that needs to happen to set up separate bundles and make lazy components. However, there are some downsides to this. If you have too many lazy components, your app will end up making several HTTP requests to fetch them – at the same time. There's no benefit to having separate bundles for components used on the same part of the app. You're better off trying to bundle components together in a way that one HTTP request is made to load what is needed on the current page.

A helpful way to think of this is to associate "pages" with bundles. If you have lazy page components, everything on that page will also be lazy yet bundled together with other components on the page. Let's build an example that demonstrates how to organize our lazy components. Let's say that your app has a couple of pages and a few features on each page. We don't necessarily want to make these features lazy if they're all going to be needed when the page loads. Here's the `App` component that shows the user a selector to pick which page to load:

```
const First = React.lazy(() => import("./First"));
const Second = React.lazy(() => import("./Second"));

function ShowComponent({ name }) {
  switch (name) {
    case "first":
      return <First />;
    case "second":
      return <Second />;
    default:
      return null;
  }
}

function App() {
  const [component, setComponent] = React.useState("");

  return (
    <>
      <label>
        Load Component:{" "}
        <select
          value={component}
          onChange={(e) => setComponent(e.target.value)}
        >
          <option value="">None</option>
          <option value="first">First</option>
          <option value="second">Second</option>
        </select>
```

```
      </label>
      <Suspense fallback="loading...">
        <ShowComponent name={component} />
      </Suspense>
    </>
  );
}
```

The `First` and `Second` components are the pages that make up our app, so we want them to be lazy components that load their bundles on demand. The `ShowComponent` component renders the appropriate page when the user changes the selector.

Next, let's look at the `First` page and see how it's composed, starting with the `First` component:

```
import One from "./One";
import Two from "./Two";
import Three from "./Three";

export default function First() {
  return (
    <>
      <One />
      <Two />
      <Three />
    </>
  );
}
```

The `First` component pulls in three components and renders them – `One`, `Two`, and `Three`. These three components will be part of the same bundle. While we could make them lazy, there would be no point, as all we would be doing is making three HTTP requests for bundles at the same time instead of one.

Now that you have a better understanding of how to map page structures of your application to bundles, let's look at another use case where we use a router component to navigate around our app.

Exploring lazy pages and routes

In the *Avoiding lazy components* section, you saw where to avoid making components lazy when there is no benefit in doing so. The same pattern can be applied when you're using `react-router` as the mechanism to navigate around your application. Let's take a look at an example. Here are the imports we'll need:

```
import {
  BrowserRouter as Router,
  Routes,
  Route,
  Link,
  Outlet,
} from "react-router-dom";
import { FadeLoader } from "react-spinners";
Next, we'll create our lazy components:
const First = React.lazy(() =>
  Promise.all([
    import("./First"),
    new Promise((resolve) => {
      setTimeout(() => {
        resolve();
      }, 3000);
    }),
  ]).then(([m]) => m)
);

const Second = React.lazy(() =>
  Promise.all([
    import("./Second"),
    new Promise((resolve) => {
      setTimeout(() => {
        resolve();
      }, 3000);
    }),
  ]).then(([m]) => m)
);
```

Finally, we have the application component that uses the two lazy components that we just declared:

```
function Layout() {
  return (
    <section>
      <nav>
        <p>
          <Link to="first">First</Link>
        </p>
        <p>
          <Link to="second">Second</Link>
        </p>
      </nav>
      <section>
        <Suspense
          fallback={<FadeLoader color={"lightblue"}
            size={150} />}
        >
          <Outlet />
        </Suspense>
      </section>
    </section>
  );
}

export default function App() {
  return (
    <Router>
      <Routes>
        <Route path="/" element={<Layout />}>
          <Route path="/first" element={<First />} />
          <Route path="/second" element={<Second />} />
        </Route>
```

```
        </Routes>
      </Router>
    );
  }
```

In the preceding code, we have two lazy page components that will be bundled separately from the rest of the app. They're using the same latency simulation technique that was introduced in the *Simulating latency* section so that we can see the fallback content as we navigate through pages by clicking on links. The fallback content in this example uses the same `FadeLoader` spinner component that was introduced in the *Working with spinner fallbacks* section.

Note that the `Suspense` component is placed beneath the navigation links. This means that the fallback content will be rendered in the spot where the page content will eventually show when it loads. The children of the `Suspense` component are the `Route` components that will render our lazy page components – for example, when the `/first` route is activated, the `First` component is rendered for the first time, triggering the bundle download.

That brings us to the end of this chapter.

Summary

This chapter was all about code splitting and bundling, which are important concepts for larger React applications. We started by looking at how code is split into bundles in your React applications by using the `import()` function. Then, we looked at the `lazy()` React API and how it helps to simplify loading bundles when components are rendered for the first time. Next, we looked more deeply at the `Suspense` component, which is used to manage content while component bundles are being fetched. The fallback property is how we specify the content to be shown while bundles are being loaded. You typically don't need more than one `Suspense` component in your app, as long as you follow a consistent pattern for bundling pages of your app.

In the next chapter, you'll learn how to use the Next.js framework to handle rendering React components on the server. The Next.js framework allows you to create pages that act as React components and can be rendered on the server and in the browser. This is an important capability for applications that need good initial page load performance – that is, all applications.

11
Server-Side React Components

Everything that you've learned so far in this book has been React code that runs in web browsers. React isn't just confined to the browser for rendering, and in this chapter, you'll learn how to render components from a Node.js server.

The first section of this chapter briefly touches upon high-level server rendering concepts. The next four sections go into more depth, teaching you how to implement the most crucial aspects of server-side rendering with React and Next.js.

In this chapter, we'll cover the following topics:

- What is isomorphic JavaScript?
- Rendering to strings
- Backend routing
- Frontend reconciliation
- Fetching data

Technical requirements

You can find the code files present in this chapter on GitHub at `https://github.com/PacktPublishing/React-and-React-Native-4th-Edition/tree/main/Chapter11`.

What is isomorphic JavaScript?

Another term for server-side rendering is **isomorphic JavaScript**. This is a fancy way of saying JavaScript code that can run in the browser and in Node.js without modification. In this section, you'll learn the basic concepts of isomorphic JavaScript before diving into the code.

The server is a render target

The beauty of React is that it's a small abstraction layer that sits on top of a rendering target. So far, the target has been the browser, but it can also be the server. The render target can be anything, just as long as the correct translation calls are implemented behind the scenes.

In the case of rendering on the server, components are rendered to strings. The server can't actually display rendered HTML; all it can do is send the rendered markup to the browser. This idea is shown in the following diagram:

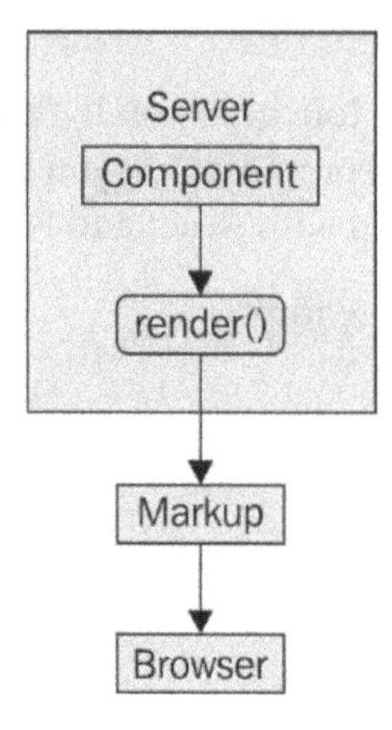

Figure 11.1 – A conceptual model of server-side rendering

It's possible to render a React component on the server and send the rendered output to the browser. The question is, why would you want to do this on the server instead of in the browser?

Initial load performance

The main motivation behind server-side rendering, for me personally, is improved performance. In particular, the initial rendering just feels faster for the user, and this translates to an overall better user experience. It doesn't matter how fast your application is once it's loaded and ready to go; it's the initial load time that leaves a lasting impression on your users.

There are three ways in which this approach results in better performance for the initial load:

- The rendering that takes place on the server is generating a string; there's no need to compute a difference or to interact with the DOM in any way. Producing a string of rendered markup is inherently faster than rendering components in the browser.
- The rendered HTML is displayed as soon as it arrives. Any JavaScript code that needs to run on the initial load is run after the user is already looking at the content.
- There are fewer network requests to fetch data from the API because these have already happened on the server, and the server, typically, has far more resources than a single client.

The following diagram illustrates these performance ideas:

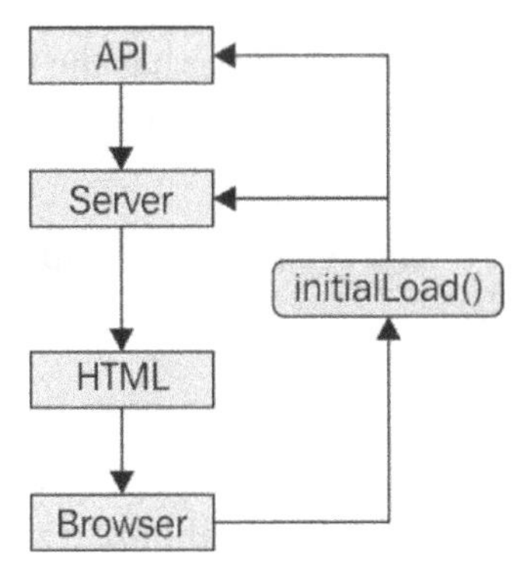

Figure 11.2 – A conceptual model of what happens during the initial load of a page that's been rendered on the server

Beyond just performance enhancements, we can share the same code between the client and the server in some cases. We'll cover this next.

Sharing code between the server and the browser

There's a good chance that your application will need to talk to API endpoints that are beyond your control; for example, an application that is composed of many different microservice endpoints. You can rarely use data from these services without modification. Rather, you have to write code that transforms data so that it's usable by your React components.

If you're rendering your components on a Node.js server, then this data transformation code will be used by both the client and the server. This is because, on the initial load, the server will need to talk to the API and, later on, the component in the browser will need to talk to the API.

It's not just about transforming the data that's returned from these services either. For example, you have to think about providing input to them as well, such as when creating or modifying resources.

The fundamental adjustment that you'll need to make as a React programmer is to assume that any given component that you implement will need to be rendered on the server. This may seem like a minor adjustment, but the devil is in the detail.

In this section, we covered the important concepts related to rendering React components on the server. You learned that React treats the server as a target to render content on, just like the browser is a target. You learned that the performance of your initial application page load can be greatly improved when the server sends content that's already been rendered. Finally, you learned that once the browser has the initial content to display, it can then use those same components that were used on the server to perform the initial render.

In the next section, we'll look at how React can render components to static HTML strings instead of DOM manipulation calls.

Rendering to strings

When React components are rendered in Node.js, they're transformed into strings of HTML output. The string content is then returned to the browser, which displays this to the user immediately. Let's look at an example.

Here is the component to render:

```
import React from "react";
import PropTypes from "prop-types";

export default function App({ items }) {
  return (
    <ul>
      {items.map(item => (
        <li key={item}>{item}</li>
      ))}
    </ul>
  );
}

App.propTypes = {
  items: PropTypes.arrayOf(PropTypes.string).isRequired
};
```

Next, let's implement the server that will render this component when the browser asks for it:

```
import * as React from "react";
import ReactDOM from "react-dom/server";
import express from "express";
import App from "./App";

const doc = (content) =>
  `
  <!doctype html>
  <html>
    <head>
      <title>Rendering to strings</title>
    </head>
    <body>
      <div id="app">${content}</div>
```

```
    </body>
  </html>
  `;

const app = express();

app.get("/", (req, res) => {
  const props = { items: ["One", "Two", "Three"] };
  const rendered = ReactDOM.renderToString(<App {...props}
    />);

  res.send(doc(rendered));
});

app.listen(8080, () => {
  console.log("Listening on 127.0.0.1:8080");
});
```

Now, if you visit `http://127.0.0.1:8080` in your browser, you'll see the rendered component content:

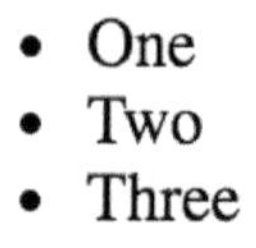

- One
- Two
- Three

Figure 11.3 – The rendered list in the browser

There are two things to pay attention to in this example. First, there's the `doc()` function. This creates the basic HTML document template with a placeholder for rendered React content. The second is the call to `renderToString()`, which is just like the `render()` call that you're used to. This is called in the server request handler and the rendered string is sent to the browser.

This section showed you that React can act similar to a template engine, by building strings as its output and using this content to serve as HTML content from the server. In the following section, we'll look at how routing works in a React application that runs on the server.

Backend routing

In the *Rendering to strings* section, you implemented a single request handler in the server that responded to requests for the root URL (/). Your application is going to need to handle more than a single route. You learned how to use the `react-router` package for routing in *Chapter 9, Handling Navigation with Routes*, but now, you're going to see how to use the same package in Node.js.

First, let's take a look at the main `App` component:

```
import { Routes, Route, Link, Outlet } from
  "react-router-dom";

import FirstHeader from "./first/FirstHeader";
import FirstContent from "./first/FirstContent";
import SecondHeader from "./second/SecondHeader";
import SecondContent from "./second/SecondContent";

function Layout() {
  return (
    <section>
      <Outlet />
    </section>
  );
}

export default function App() {
  return (
    <Routes>
      <Route path="/" element={<Layout />}>
        <Route
          index
          element={
            <>
              <h1>App</h1>
              <ul>
                <li>
```

```
            <Link to="first">First</Link>
          </li>
          <li>
            <Link to="second">Second</Link>
          </li>
        </ul>
      </>
    }
  />
  <Route
    path="/first"
    element={
      <>
        <header>
          <FirstHeader />
        </header>
        <main>
          <FirstContent />
        </main>
      </>
    }
  />
  <Route
    path="/second"
    element={
      <>
        <header>
          <SecondHeader />
        </header>
        <main>
          <SecondContent />
        </main>
      </>
    }
  />
</Route>
```

```
      </Routes>
    );
}
```

There are three routes that this application handles:

- `/`: The home page
- `/first`: The first page of content
- `/second`: The second page of content

The `App` content is divided into `<header>` and `<main>` elements. In each of these sections, there is a `<Route>` component that handles the appropriate content. For example, the main content for the `/` route is handled by a `render()` function that renders links to `/first` and `/second`.

This component will work fine on the client, but will it work on the server? Let's implement that now:

```
import React from "react";
import { renderToString } from "react-dom/server";
import { StaticRouter } from "react-router-dom/server";
import express from "express";

import App from "./App";

const app = express();

app.get("/*", (req, res) => {
  const html = renderToString(
    <StaticRouter location={req.url}>
      <App />
    </StaticRouter>
  );
  res.write('
      <!doctype html>
      <div id="app">${html}</div>
    ');
  res.end();
```

```
});

app.listen(8080, () => {
  console.log("Listening on 127.0.0.1:8080");
});
```

Now you have both frontend and backend routing! How does this work exactly? Let's start with the request handler path. This has changed so that it's now a wildcard (`/*`). Now, this handler is called for every request.

On the server, the `<StaticRouter>` component is used instead of the `<BrowserRouter>` component. The `<App>` component is the child, which means that the `<Route>` components within it will be passed data from `<StaticRouter>`. This is how `<App>` knows to render the correct content based on the URL. The resulting HTML value that results from calling `renderToString()` can then be used as part of the document that's sent to the browser as a response.

Now your application is starting to look like a real end-to-end React rendering solution. This is what the server renders if you hit the root URL, `/`:

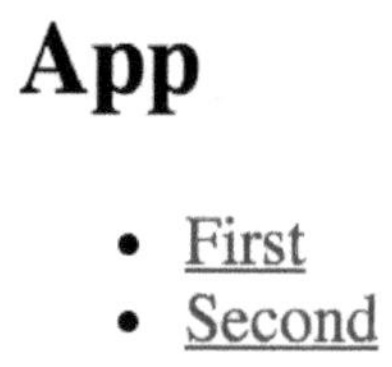

App

- First
- Second

Figure 11.4 – The output of the index page

If you hit the `/second` URL, the Node.js server will render the correct component:

Second

Second

Figure 11.5 – The output of the second page

If you navigate from the main page to the first page, the request goes back to the server. We need to figure out how to get the frontend code to the browser so that it can take over after the initial render.

In this section, you learned that `react-router` routes work similarly to how they would work in a browser-based React app. In the next section, we'll make sure that your components can work both on the server and in the browser.

Frontend reconciliation

The only thing that was missing from the last example was the client's JavaScript code. The user wants to use the application, and the server needs to deliver the client's code bundle. How would this work? Routing has to work in the browser and on the server, without modifying the routes. In other words, the server handles routing in the initial request and then the browser takes over as the user starts clicking on things and moving around in the application.

Let's create the `index.js` module for this example:

```
import React from "react";
import { hydrate } from "react-dom";
import App from "./App";

hydrate(<App />, document.getElementById("root"));
```

This looks like most other `index.js` files that you've seen so far in this book. You render the `<App>` component in the root element in the HTML document. In this case, you're using the `hydrate()` function instead of the `render()` function. The two functions have the same end result—rendered JSX content in the browser window. The `hydrate()` function is different because it expects rendered component content to already be in place. This means that it will perform less work because it will assume that the markup is correct and doesn't need to be updated on the initial render.

Only in development mode will React examine the entire DOM tree of the server-rendered content to make sure that the correct content is displayed. If there's a mismatch between the existing content and the output of the React components, you'll see warnings that show you where these mismatches happened, so that you can go and fix them.

Here is the `App` component that your app will render in the browser and on the Node.js server:

```
import React, { useState } from "react";

export default function App() {
  const [clicks, setClicks] = useState(0);

  return (
    <section>
      <header>
        <h1>Hydrating The Client</h1>
```

```
        </header>
        <main>
          <p>Clicks {clicks}</p>
          <button onClick={() => setClicks(clicks + 1)}>Click
            Me</button>
        </main>
      </section>
    );
}
```

The component renders a button that, when clicked, will update the `clicks` state. This state is rendered in a label above the button. When this component is rendered on the server, the default `clicks` value of `0` is used, and the `onClick` handler is ignored since it's just rendering static markup. Let's take a look at the server code next:

```
import fs from "fs";
import React from "react";
import { renderToString } from "react-dom/server";
import express from "express";
import App from "./App";

const app = express();
const doc = fs.readFileSync("./build/index.html");

app.use(express.static("./build", { index: false }));

app.get("/*", (req, res) => {
  const context = {};
  const html = renderToString(<App />);

  if (context.url) {
    res.writeHead(301, {
      Location: context.url
    });
    res.end();
  } else {
    res.write(
```

```
        doc.toString().replace('<div id="root">',
          '<div id="root">${html}')
      );
      res.end();
    }
  });

  app.listen(8080, () => {
    console.log("Listening on 127.0.0.1:8080");
  });
```

Let's walk through the preceding code and see what's going on:

```
const doc = fs.readFileSync("./build/index.html");
```

The previous line of code reads the `index.html` file that's created by your React build tool, such as `create- react-app/react-scripts`, and stores it in doc.

Next, we added this:

```
app.use(express.static("./build", { index: false }));
```

This tells the Express server to serve files under `./build` as static files, except for `index.html`.

Instead, you're going to write a handler that responds to requests for the root of the site:

```
app.get("/*", (req, res) => {
  const context = {};
  const html = renderToString(<App />);

  if (context.url) {
    res.writeHead(301, {
      Location: context.url
    });
    res.end();
  } else {
    res.write(
      doc.toString().replace('<div id="root">',
        '<div id="root">${html}')
```

```
    );
    res.end();
  }
});
```

This is where the HTML constant is populated with the rendered React content. Then, it gets interpolated into the HTML string using `replace()` and is sent as the response. Because you've used the `index.html` file based on your build, it contains a link to the bundled React app that will run when loaded in the browser.

In this section, you learned how to share the same components that render content on the server with your application that runs in the browser. In the next section, you'll learn how to leverage Next.js to fetch data that React components on the server need.

Fetching data

What if one of your components needs to fetch API data before it can fully render its content? This presents a challenge for rendering on the server because there's no easy way to define a component that knows when to fetch data on the server and in the browser.

This is where a minimal framework such as Next.js comes into play. Next.js treats server rendering and browser rendering as equals. This means that the headache of fetching data for your components is abstracted—you can use the same code in the browser and on the server.

To handle routing, Next.js uses the concept of pages. A page is a JavaScript module that exports a React component. The rendered content of the component turns into the page content. Here's what the pages directory looks like:

- `pages`
  - `first.js`
  - `index.js`
  - `second.js`

The `index.js` module is the root page of the app; Next.js knows this based on the filename. Here's what the source looks like:

```
import Layout from "../components/MyLayout.js";
export default function Index() {
  return (
    <Layout>
```

```
        <p>Fetching component data on the server and on the
          client...</p>
      </Layout>
    );
}
```

This page uses a `<Layout>` component to ensure that common components are rendered without the need to duplicate code. Here's what the page looks like when rendered:

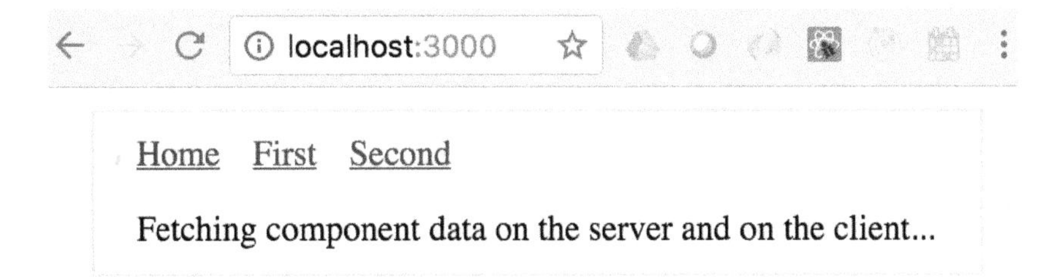
localhost:3000
Home First Second
Fetching component data on the server and on the client...

Figure 11.6 – The index page of the Next.js app

In addition to the paragraph, you have the overall application layout, including the navigation links to other pages. Here's what the `Layout` source looks like:

```
import Header from "./Header";

const layoutStyle = {
  margin: 20,
  padding: 20,
  border: "1px solid #DDD"
};

export default function Layout(props) {
  return (
    <div style={layoutStyle}>
      <Header />
      {props.children}
    </div>
  );
}
```

The `Layout` component renders a `Header` component and `props.children`. The `children` property is the value that you pass to the `Layout` component in your pages.

Let's take a look at the `Header` component now:

```
import Link from "next/link";

const linkStyle = {
  marginRight: 15
};

export default function Header() {
  return (
    <div>
      <Link href="/">
        <a style={linkStyle}>Home</a>
      </Link>
      <Link href="/first">
        <a style={linkStyle}>First</a>
      </Link>
      <Link href="/second">
        <a style={linkStyle}>Second</a>
      </Link>
    </div>
  );
}
```

The `Link` component used here comes from Next.js. This is so that the links work as expected with the routing that Next.js sets up automatically.

Now, let's look at a page that has data-fetching requirements—`pages/first.js`:

```
import Layout from "../components/MyLayout.js";
import { fetchFirstItems } from "../api";

export default function First({ items }) {
  return (
```

```
    <Layout>
      {items.map(item => (
        <li key={item}>{item}</li>
      ))}
    </Layout>
  );
}

First.getInitialProps = async () => {
  const res = await fetchFirstItems();
  const items = await res.json();

  return { items };
};
```

The `fetch()` function that's used to fetch data comes from the `isomorphic-unfetch` package. This version of `fetch()` works on the server and in the browser. There's no need for you to check anything. Once again, the `Layout` component is used to wrap the page content for consistency with other pages.

The `getInitialProps()` function is how Next.js fetches data—in the browser and on the server. This is an async function, meaning that you can take as long as you need to fetch data for the component properties and Next.js will make sure not to render any markup until the data is ready.

Let's take a look at the `fetchFirstItems()` API function:

```
export default function fetchFirstItems() {
  return new Promise(resolve =>
    setTimeout(() => {
      resolve({
        json: () => Promise.resolve(["One", "Two",
          "Three"])
      });
    }, 1000)
  );
}
```

This function is mimicking API behavior by returning a promise that's resolved after 1 second with data for the component. If you navigate to `/first`, you'll see the following after 1 second:

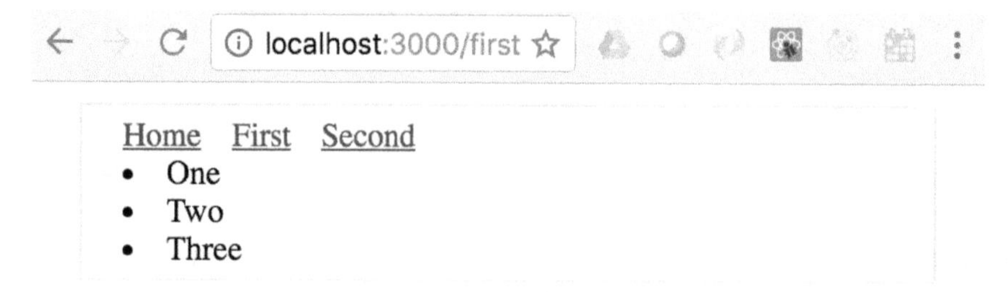

Figure 11.7 – The first page of the Next.js app

By clicking on the **First** link, you caused the `getInitialProps()` function to be called in the browser since the app has already been delivered. If you reload the page while at `/first`, you'll trigger `getInitialProps()` to be called on the server since this is the page that Next.js is handling on the server.

Summary

In this chapter, you learned that React can be rendered on the server, in addition to the client. There are several reasons for doing this, such as sharing common code between the frontend and the backend. The main advantage of server-side rendering is the performance boost that you get on the initial page load. This translates to a better user experience and, therefore, a better product.

Then, you progressively improved a server-side React application, starting with a single-page render. You were also introduced to routing, client-side reconciliation, and component data fetching to produce a complete backend rendering solution using Next.js.

In the next chapter, you'll learn how to implement modern React UI components using Material UI.

12
User Interface Framework Components

If you're using React to build a **user interface** (**UI**) for your application, you probably aren't planning on creating your own UI library too. There are lots of React UI component libraries available to choose from, and there's no wrong choice, as long as the components make your life simpler.

This chapter will introduce you to the Material-UI React library. Here are the specific topics that we'll cover:

- Layout and organization
- Using navigation components
- Collecting user input
- Working with styles and themes

Technical requirements

You can find the code files present in this chapter on GitHub at `https://github.com/PacktPublishing/React-and-React-Native-4th-Edition/tree/main/Chapter12`.

Layout and organization

Material-UI provides us with several components that help us control the overall layout of our applications and organize the other UI components without each layout. This section will demonstrate how to use containers and grids.

Using containers

Often, when you're trying to lay components out on your page, the horizontal layout is the most difficult part to get right. The `Container` component from Material-UI is a simple but powerful layout tool. It helps to control the horizontal width of its children. Let's look at an example to see what's possible:

```
import "typeface-roboto";
import React, { Fragment } from "react";
import Typography from "@mui/material/Typography";
import Container from "@mui/material/Container";

export default function App() {
  const textStyle = {
    backgroundColor: "#cfe8fc",
    margin: 5,
    textAlign: "center",
  };

  return (
    <Fragment>
      <Container maxWidth="sm">
        <Typography style={textStyle}>sm</Typography>
      </Container>
      <Container maxWidth="md">
        <Typography style={textStyle}>md</Typography>
      </Container>
```

```
        <Container maxWidth="lg">
          <Typography style={textStyle}>lg</Typography>
        </Container>
      </Fragment>
    );
  }
```

This example has three `Container` components, each of which wraps a `Typography` component. The `Typography` component is used to render text in Material-UI applications. Each `Container` component used in this example takes a `maxWidth` property. It accepts a breakpoint string value. These breakpoints represent common screen sizes. This example uses small (`sm`), medium (`md`), and large (`lg`). When the screen reaches these breakpoint sizes, the container width will stop growing. Here's what the page looks like when the width is smaller than the `sm` breakpoint:

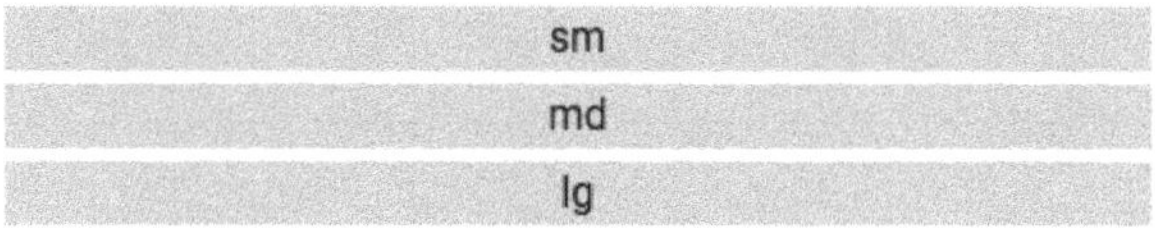

sm
md
lg

Figure 12.1 – The sm breakpoint

Now, if we were to resize the screen so that it was larger than the `md` breakpoint, but smaller than the `lg` breakpoint, here is what it would look like:

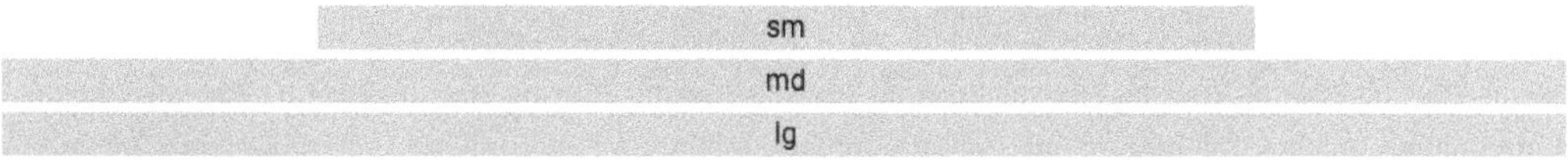

sm
md
lg

Figure 12.2 – The lg breakpoint

Notice how the first container stays at a fixed width now that we've exceeded its `maxWidth` breakpoint. The `md` and `lg` containers just keep growing along with the screen until their breakpoints have been passed. Let's see what these `Container` components look like when the screen width surpasses all breakpoints:

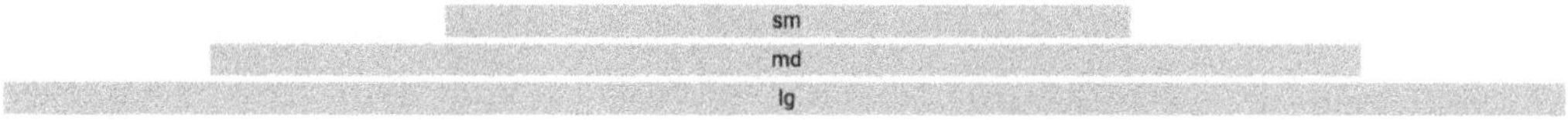

sm
md
lg

Figure 12.3 – All breakpoints

The `Container` component gives you control over how your page elements grow horizontally. They're also responsive, so your layouts will be updated as the screen dimensions change. While it is helpful, we can only do so much with horizontal layouts. In the next section, we'll look at using Material-UI components to build more complex and responsive layouts.

Building responsive grid layouts

Material-UI has a `Grid` component that we can use to compose responsive complex layouts. At a high level, a `Grid` component can be either a container or an item within a container. By combining these two roles, we can achieve any type of layout for our app. To get familiar with Material-UI grid layouts, let's put together an example that uses a common layout pattern that we'll find in many web applications. Here is what the result looks like:

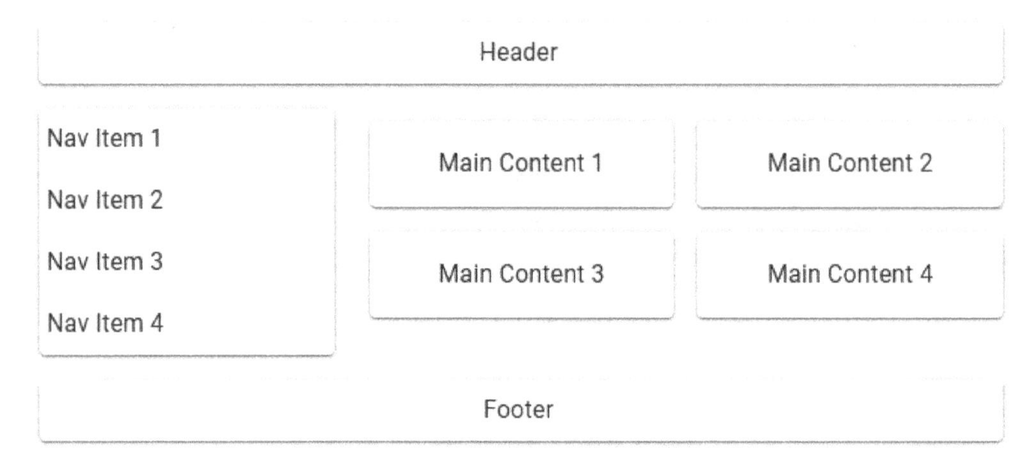

Figure 12.4 – A sample responsive grid layout

As you can see, this layout has familiar sections that are typical in many web apps. This is just an example layout; you can use the `Grid` component to build any type of layout you can imagine. Let's look at the code that created this layout:

```
import "typeface-roboto";
import React from "react";
import Paper from "@mui/material/Paper";
import Grid from "@mui/material/Grid";
import Typography from "@mui/material/Typography";

const headerFooterStyle = {
```

```
  padding: 8,
  textAlign: "center",
};
const mainStyle = {
  padding: 16,
  textAlign: "center",
};
const navStyle = { marginLeft: 5 };

export default function App() {
  return (
    <div style={{ flexGrow: 1 }}>
      <Grid container spacing={3}>
        <Grid item xs={12}>
          <Paper style={headerFooterStyle}>
            <Typography>Header</Typography>
          </Paper>
        </Grid>
        <Grid item xs={4}>
          <Paper>
            <Grid container spacing={2} direction="column">
              <Grid item xs={12}>
                <Typography style={navStyle}>Nav Item
                  1</Typography>
              </Grid>
              <Grid item xs={12}>
                <Typography style={navStyle}>Nav Item
                  2</Typography>
              </Grid>
              <Grid item xs={12}>
                <Typography style={navStyle}>Nav Item
                  3</Typography>
              </Grid>
              <Grid item xs={12}>
                <Typography style={navStyle}>Nav Item
                  4</Typography>
```

```
            </Grid>
          </Grid>
        </Paper>
      </Grid>
      <Grid item xs={8}>
        <Grid container spacing={2}>
          <Grid item xs={6}>
            <Paper style={mainStyle}>
              <Typography>Main Content 1</Typography>
            </Paper>
          </Grid>
          <Grid item xs={6}>
            ...
          </Grid>
          <Grid item xs={6}>
            ...
          </Grid>
          <Grid item xs={6}>
            ...
          </Grid>
        </Grid>
      </Grid>
      <Grid item xs={12}>
        <Paper style={headerFooterStyle}>
          <Typography>Footer</Typography>
        </Paper>
      </Grid>
    </Grid>
  </div>
);
}
```

There are a couple of places where I've replaced repetitive code with `...`. In these cases, the code that was removed was just a repeat of the `Grid` component that came before it. Now, let's break down how the sections in this layout are created. We'll start with the header section:

```
<Grid item xs={12}>
  <Paper style={headerFooterStyle}>
    <Typography>Header</Typography>
  </Paper>
</Grid>
```

The `xs` breakpoint property value of `12` means that the header will always span the entire width of the screen since 12 is the highest value you can use here. Next, let's look at the navigation items:

```
<Grid item xs={4}>
  <Paper>
    <Grid container spacing={2} direction="column">
      <Grid item xs={12}>
        <Typography style={navStyle}>Nav Item
          1</Typography>
      </Grid>
      <Grid item xs={12}>
        <Typography style={navStyle}>Nav Item
          2</Typography>
      </Grid>
      <Grid item xs={12}>
        <Typography style={navStyle}>Nav Item
          3</Typography>
      </Grid>
      <Grid item xs={12}>
        <Typography style={navStyle}>Nav Item
          4</Typography>
      </Grid>
    </Grid>
  </Paper>
</Grid>
```

In the navigation section, we have a grid container nested inside of a grid item. It's common to nest grids like this, and the more complex the layout, the more levels of nested grids that you'll require. You'll notice that the direction property value of the column used in the navigation section makes the navigation items flow vertically instead of the horizontal default. Next, we'll look at the main content section:

```
<Grid item xs={8}>
  <Grid container spacing={2}>
    <Grid item xs={6}>
      <Paper style={mainStyle}>
        <Typography>Main Content 1</Typography>
      </Paper>
    </Grid>
    <Grid item xs={6}>
      <Paper style={mainStyle}>
        <Typography>Main Content 2</Typography>
      </Paper>
    </Grid>
    <Grid item xs={6}>
      <Paper style={mainStyle}>
        <Typography>Main Content 3</Typography>
      </Paper>
    </Grid>
    <Grid item xs={6}>
      <Paper style={mainStyle}>
        <Typography>Main Content 4</Typography>
      </Paper>
    </Grid>
  </Grid>
</Grid>
```

The main content section follows the same approach as the navigation section—it uses a nested grid container for subsections. The `xs` breakpoint value of `6` used by each of the `Grid` subsection components determines how wide each of them is and how they flow on the page. Since the value is 6, they take up half of the available space in the main section.

Also, you can see that the `xs` breakpoint value for the main section is `8`. The `xs` value for the navigation section is `4`; these two numbers add up to 12, meaning that, together, they use the full width of the screen.

In this section, you were introduced to what Material-UI has to offer in the way of layouts. You can use the `Container` component to control the width of sections and how they change in response to screen dimension changes. You then learned that the `Grid` component is used to put together more complex grid layouts. In the following section, we'll look at some of the navigational components found in Material-UI.

Using navigation components

Once we have an idea of how the layout of our application is going to look and work, we can start to think about the navigation. This is an important piece of our UI because it's how the user gets around the application, and it will be used frequently. In this section, we'll learn about two of the navigational components offered by Material-UI.

Navigating with drawers

The `Drawer` component, just like a physical drawer, slides open to reveal content that is easily accessed. When we're finished, the drawer closes again. This works well for navigation because it stays out of the way, allowing more space on the screen for the active task that the user is engaged with. Let's look at an example, starting with the `App` component:

```
import First from "./First";
import Second from "./Second";
import Third from "./Third";

export default function App({ links }) {
  const [open, setOpen] = useState(false);

  function toggleDrawer({ type, key }) {
    if (type === "keydown" && (key === "Tab" || key ===
      "Shift")) {
      return;
    }

    setOpen(!open);
  }

  return (
```

```
<Router>
  <Button onClick={toggleDrawer}>Open Nav</Button>
  <section>
    <Route path="/first" component={First} />
    <Route path="/second" component={Second} />
    <Route path="/third" component={Third} />
  </section>
  <Drawer open={open} onClose={toggleDrawer}>
    <div
      style={{ width: 250 }}
      role="presentation"
      onClick={toggleDrawer}
      onKeyDown={toggleDrawer}
    >
      <List>
        {links.map((link) => (
          <ListItem button key={link.url}
            component={Link} to={link.url}>
            <Switch>
              <Route
                exact
                path={link.url}
                render={() => (
                  <ListItemText
                    primary={link.name}
                    primaryTypographyProps={{ color:
                      "primary" }}
                  />
                )}
              />
              <Route
                path="/"
                render={() => <ListItemText
                  primary={link.name} />}
              />
```

```
                </Switch>
              </ListItem>
            ))}
          </List>
        </div>
      </Drawer>
    </Router>
  );
}
```

Let's look at what's happening here. Everything that this component renders is within the `Router` component because the items in the drawer are links to routes:

```
<section>
  <Route path="/first" component={First} />
  <Route path="/second" component={Second} />
  <Route path="/third" component={Third} />
</section>
```

The `First`, `Second`, and `Third` components are used to render the main application content when the user clicks on a link in the drawer. The drawer itself is opened when the **Open Nav** button is clicked. Let's take a closer look at the state that's used to control this:

```
const [open, setOpen] = useState(false);

function toggleDrawer({ type, key }) {
  if (type === "keydown" && (key === "Tab" ||
    key === "Shift")) {
    return;
  }

  setOpen(!open);
}
```

The open state controls the visibility of the drawer. The `onClose` property of the `Drawer` component calls this function, too, meaning that the drawer closes when any of the links within it are activated. Next, let's look at how the links within the drawer are generated:

```
<List>
  {links.map((link) => (
    <ListItem button key={link.url} component={Link}
      to={link.url}>
      <Switch>
        <Route
          exact
          path={link.url}
          render={() => (
            <ListItemText
              primary={link.name}
              primaryTypographyProps={{ color: "primary" }}
            />
          )}
        />
        <Route
          path="/"
          render={() => <ListItemText primary={link.name}
            />}
        />
      </Switch>
    </ListItem>
  ))}
</List>
```

The items that are displayed in a `Drawer` component are actually list items, as you can see here. The `links` property that is passed to `App` has all the link objects with the `url` and `name` properties. Each item in the `items` array is mapped to the `ListItem` component, which uses the `Link` component. Within `ListItem`, we have the `Route` component that generates the link text by rendering a `ListItemText` component. There are two `Route` components within a `Switch` component. The reason is so that we can style the list item differently if it matches the current path. Finally, let's look at the default value for the `links` property:

```
App.defaultProps = {
  links: [
    { url: "/first", name: "First Page" },
    { url: "/second", name: "Second Page" },
    { url: "/third", name: "Third Page" },
  ],
};
```

Here's what the drawer looks like when it's opened after the screen first loads:

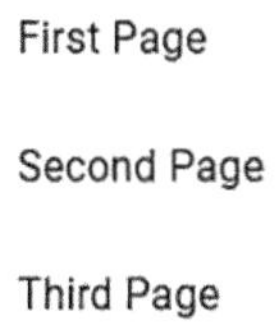

First Page

Second Page

Third Page

Figure 12.5 – A drawer showing links to our pages

Try clicking on the **First Page** link. The drawer closes and renders the content of the `/first` route. Then, when you open the drawer again, you'll notice that the **First Page** link is rendered as the active link:

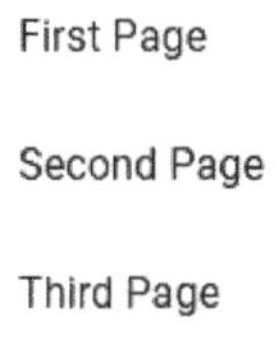

First Page

Second Page

Third Page

Figure 12.6 – The first page link is styled as the active link in the drawer

In this section, you learned how to use the `Drawer` component as the main navigation for your application. In the following section, we'll look at the `Tabs` component.

Navigating with tabs

Tabs are another common navigation pattern found in modern web apps. The `Material-UI Tabs` component lets us use tabs as links and hook them up to a router. Let's look at an example of how to do this. Here is the `App` component:

```
const tabContentStyle = {
  padding: 16,
};

function TabContainer({ value }) {
  return (
    <AppBar position="static">
      <Tabs value={value}>
        <Tab label="Item One" component={Link} to="/" />
        <Tab label="Item Two" component={Link} to="/page2"
          />
        <Tab label="Item Three" component={Link}
          to="/page3" />
      </Tabs>
    </AppBar>
  );
}

export default function App() {
  return (
    <Router>
      <Route
        exact
        path="/"
        render={() => (
          <>
            <TabContainer value={0} />
            <Typography component="div"
```

```
                style={tabContentStyle}>
                Item One
              </Typography>
            </>
          )}
        />
        <Route
          exact
          path="/page2"
          render={() => (
            <>
              <TabContainer value={1} />
              <Typography component="div"
                style={tabContentStyle}>
                Item Two
              </Typography>
            </>
          )}
        />
      </Router>
    );
}
```

In the interest of space, I've left out the `Route` component for `/page3`; it follows the same pattern as `/page2`. The `Tabs` and `Tab` components from Material-UI don't actually render any content underneath the selected tab. It's up to us to provide the content as the `Tabs` component only looks after showing the tabs and marking one of them as selected. This example aims to have the `Tab` components use `Link` components that link to content rendered by routes. Let's now take a closer look at the `TabContainer` component:

```
function TabContainer({ value }) {
  return (
    <AppBar position="static">
      <Tabs value={value}>
        <Tab label="Item One" component={Link} to="/" />
        <Tab label="Item Two" component={Link} to="/page2"
          />
        <Tab label="Item Three" component={Link}
```

```
          to="/page3" />
        </Tabs>
      </AppBar>
    );
}
```

Here, we're wrapping the `Tabs` component with the `AppBar` component, meaning that the tabs appear like they're part of the bar across the top of the UI. Each `Tab` component uses the `Link` component so that, when it is clicked, the router is activated with the route specified in the `to` property. The `TabContainer` component is then used as a child component inside our `Route` components. This way, the route knows which value property to pass—this determines the active tab.

Here's what the page looks like when it first loads:

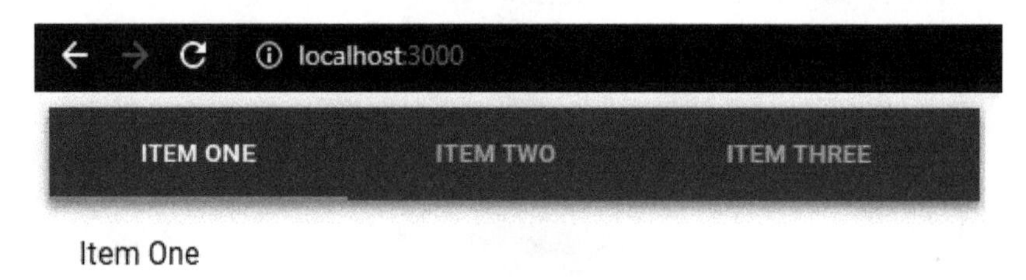

Figure 12.7 – Tabs with the first item active

If you click on the **ITEM TWO** tab, the URL will update, the active tab will change, and the page content below the tabs will change:

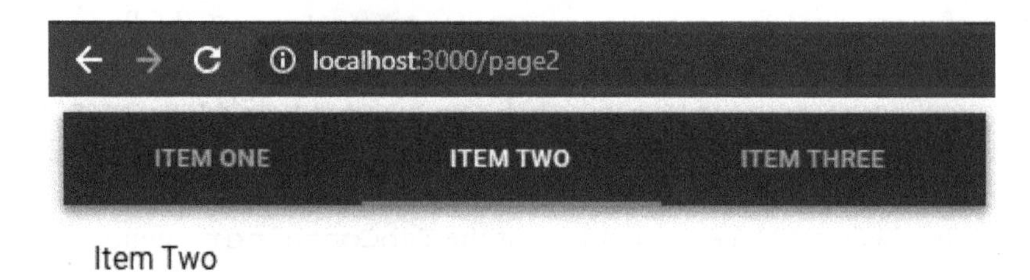

Figure 12.8 – Tabs with the second item active

In this section, you learned about two of the navigation approaches that you can use in your Material-UI application. The first is to use a drawer that is only displayed when the user needs to access navigational links. The second is to use tabs that are always visible. In the following section, you'll learn about collecting input from users.

Collecting user input

Collecting input from users can be difficult. There are many nuanced things about every field that we need to consider if we plan on getting the user experience right. Thankfully, the form components available in Material-UI take care of a lot of usability concerns for us. In this section, you'll get a brief sampling of the input controls that you can use.

Checkboxes and radio buttons

Checkboxes are useful for collecting true/false answers from users, while radio buttons are useful for getting the user to select an option from a short number of choices. Let's take a look at an example of these components in Material-UI:

```
import "typeface-roboto";
import React from "react";
import Checkbox from "@mui/material/Checkbox";
import Radio from "@mui/material/Radio";
import RadioGroup from "@mui/material/RadioGroup";
import FormControlLabel from
  "@mui/material/FormControlLabel";
import FormControl from "@mui/material/FormControl";
import FormLabel from "@mui/material/FormLabel";

export default function Checkboxes() {
  const [checkbox, setCheckbox] = React.useState(false);
  const [radio, setRadio] = React.useState("First");

  return (
    <div>
      <FormControlLabel
        label={'Checkbox ${checkbox ? "(checked)" : ""}'}
        control={
          <Checkbox
            checked={checkbox}
            onChange={() => setCheckbox(!checkbox)}
          />
        }
      />
```

```
        <FormControl component="fieldset">
          <FormLabel component="legend">{radio}</FormLabel>
          <RadioGroup value={radio} onChange={(e) =>
            setRadio(e.target.value)}>
            <FormControlLabel value="First" label="First"
              control={<Radio />} />
            <FormControlLabel value="Second" label="Second"
              control={<Radio />} />
            <FormControlLabel value="Third" label="Third"
              control={<Radio />} />
          </RadioGroup>
        </FormControl>
      </div>
    );
}
```

This example has two pieces of state. The checkbox state controls the value of the `Checkbox` component, while the radio value controls the state of the `RadioGroup` component. The checkbox state is passed to the `checked` property of the `Checkbox` component, while the radio state is passed to the `value` property of the `RadioGroup` component. Both components have `onChange` handlers that call their respective state setter functions: `setCheckbox()` and `setRadio()`. You'll notice that many other Material-UI components are involved in the display of these controls. For example, the label for the checkbox is displayed using the `FormControlLabel` component, while the radio control uses a `FormControl` component and a `FormLabel` component.

Here is what the two input controls look like:

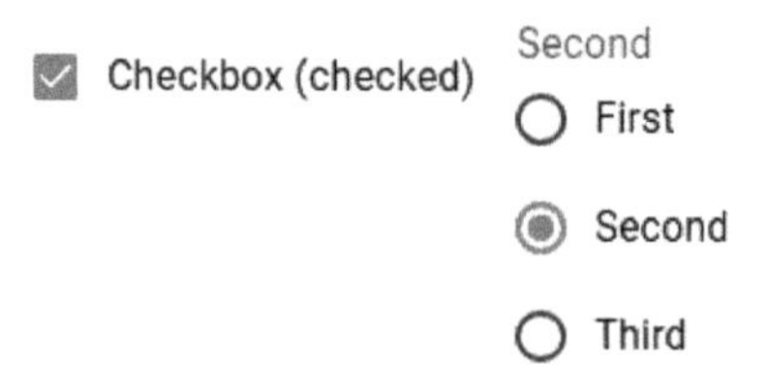

Figure 12.9 – A checkbox and a radio group

The labels for both of these controls are updated to reflect the state of the component as they change. The checkbox labels show whether the checkbox is checked, and the radio labels show the currently selected value. In the next section, we'll look at text inputs and select components.

Text inputs and select inputs

Text fields allow our users to enter text, while selects allow them to choose from several options. The difference between selects and radio buttons is that selects require less space on the screen since the options are only displayed when the user opens the option menu.

Let's look at a `Select` component now:

```
import React, { useState } from "react";
import InputLabel from "@mui/material/InputLabel";
import MenuItem from "@mui/material/MenuItem";
import FormControl from "@mui/material/FormControl";
import Select from "@mui/material/Select";

export default function MySelect() {
  const [value, setValue] = useState("first");

  return (
    <FormControl>
      <InputLabel htmlFor="my-select">My
        Select</InputLabel>
      <Select
        value={value}
        onChange={(e) => setValue(e.target.value)}
        inputProps={{ id: "my-select" }}
      >
        <MenuItem value="first">First</MenuItem>
        <MenuItem value="second">Second</MenuItem>
        <MenuItem value="third">Third</MenuItem>
      </Select>
    </FormControl>
  );
}
```

The value state used in this example controls the selected value in the `Select` component. When the user changes their selection, the `setValue()` function changes the value. The `MenuItem` component is used to specify the available options in the `select` field; the `value` property is set as the value state when a given item is selected. Here's what the select looks like when the menu is displayed:

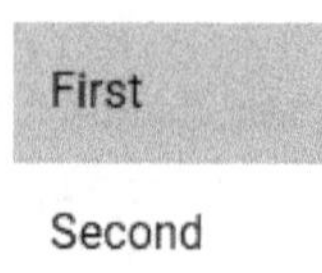

Figure 12.10 – A menu with the first item active

Next, let's look at a `TextField` component example:

```
import React, { useState } from "react";
import TextField from "@mui/material/TextField";

export default function MyTextInput() {
  const [value, setValue] = useState("");

  return (
    <TextField
      label="Name"
      value={value}
      onChange={(e) => setValue(e.target.value)}
      margin="normal"
    />
  );
}
```

The value state controls the value of the text input and changes as the user types. Here's what the text field looks like:

Name

My Name

Figure 12.11 – A text field with user-provided text

Unlike other form control components, the `TextField` component doesn't require several other supporting components. Everything that we need can be specified via properties. In the next section, we'll look at the `Button` component.

Working with buttons

Material-UI buttons are very similar to HTML button elements. The difference is that they're React components that work well with other aspects of Material-UI, such as theming and layout. Let's look at an example that renders different styles of buttons:

```
const buttonStyle = { margin: 10 };

function toggleColor(setter, value) {
  setter(value === "default" ? "primary" : "default");
}

export default function App() {
  const [contained, setContained] = useState("default");
  const [text, setText] = useState("default");
  const [outlined, setOutlined] = useState("default");
  const [icon, setIcon] = useState("default");

  return (
    <Grid container>
      <Grid
        item
        component={Button}
        variant="contained"
        style={buttonStyle}
        color={contained}
        onClick={() => toggleColor(setContained,
          contained)}
      >
        Contained
      </Grid>
      <Grid
```

```
          item
          component={Button}
          style={buttonStyle}
          color={text}
          onClick={() => toggleColor(setText, text)}
        >
          Text
        </Grid>
        <Grid
          item
          component={Button}
          variant="outlined"
          style={buttonStyle}
          color={outlined}
          onClick={() => toggleColor(setOutlined, outlined)}
        >
          Outlined
        </Grid>
        <Grid
          item
          component={IconButton}
          style={buttonStyle}
          color={icon}
          onClick={() => toggleColor(setIcon, icon)}
        >
          <AndroidIcon />
        </Grid>
      </Grid>
  );
}
```

This example renders four different button styles. We're using the `Grid` component to render the row of buttons. Instead of rendering buttons as children of the `Grid` item components, we're setting the `component` property value to `Button` and `IconButton`. This way, we can pass button properties directly to `Grid`. Each button has its own color state, initially set to default. When the buttons are clicked on, the state toggles to primary.

Here's what the buttons look like when they're first rendered:

Figure 12.12 – Four styles of Material UI buttons

And here's what the buttons look like when they've each been clicked on:

Figure 12.13 – What the buttons look like after they've been clicked on

In this section, you learned about some of the user input controls available in Material-UI. Checkboxes and radio buttons are useful when the user needs to turn something on or off or choose an option. Text inputs are necessary when the user needs to type in some text, while selects are useful when you have a list of options to choose from but limited space to display those options. Finally, you learned that Material-UI has several styles of buttons that can be used when the user needs to initiate an action. In the following section, we'll look at how styles and themes work in Material-UI.

Working with styles and themes

Included with Material-UI are systems for extending the styles of UI components and extending theme styles that are applied to all components. In this section, you'll learn about using both these systems.

Making styles

Material-UI comes with a `styled()` function that can be used to create styled components based on JavaScript objects. The return value of this function is a new component with the new styles applied.

Let's take a closer look at this approach:

```
import "typeface-roboto";
import React from "react";
import { styled } from "@mui/material/styles";
import Button from "@mui/material/Button";

const StyledButton = styled(Button)(({ theme }) => ({
```

```
  "&.MuiButton-root": { margin: theme.spacing(1) },
  "&.MuiButton-contained": { borderRadius:
    theme.shape.borderRadius + 2 },
  "&.MuiButton-sizeSmall": { fontWeight:
    theme.typography.fontWeightLight },
}));

export default function App() {
  return (
    <>
      <StyledButton>First</StyledButton>
      <StyledButton
        variant="contained">Second</StyledButton>
      <StyledButton size="small" variant="outlined">
        Third
      </StyledButton>
    </>
  );
}
```

The names used in this style (`MuiButton-root`, `MuiButton-contained`, and `MuiButton-sizeSmall`) aren't something that we came up with. These are part of the Button CSS API. The root style is applied to all buttons, so, in this example, all three buttons will have the margin value that we've applied here. The contained style is applied to buttons that use the `contained` variant. The `sizeSmall` style is applied to buttons that have a small value for the `size` property.

Here's what the custom button styles look like:

Figure 12.14 – Buttons using customized styles

Now that you know how to change the look and feel of individual components, it's time to think about customizing the look and feel of the application as a whole.

Customizing themes

Material-UI comes with a default theme. We can use this as the starting point to create our own theme. There are two main steps to creating a new theme in Material-UI:

1. Use the `createTheme()` function to customize the default theme settings and return a new theme object.
2. Use the `ThemeProvider` component to wrap our application so that the appropriate theme is applied.

Let's look at how this process works in practice:

```
import "typeface-roboto";
import React from "react";
import { createTheme, ThemeProvider } from
  "@mui/material/styles";
import Menu from "@mui/material/Menu";
import MenuItem from "@mui/material/MenuItem";

const theme = createTheme({
  typography: {
    fontSize: 11,
  },
  overrides: {
    MuiMenuItem: {
      root: {
        marginLeft: 15,
        marginRight: 15,
      },
    },
  },
});

export default function App() {
  return (
    <ThemeProvider theme={theme}>
      <Menu anchorEl={document.body} open={true}>
        <MenuItem>First Item</MenuItem>
        <MenuItem>Second Item</MenuItem>
```

```
            <MenuItem>Third Item</MenuItem>
          </Menu>
        </ThemeProvider>
    );
  }
```

The custom theme that we've created here does two things:

- It changes the default font size for all components to `11`.
- It updates the left and right margin values for the `MenuItem` components.

Many values can be set in a Material-UI theme; refer to the customization documentation for more. The `overrides` section is used for component-specific customizations. This is useful when you need to style for every instance of a component in your application.

Summary

This chapter was a very brief introduction to Material-UI, the most popular React UI framework. We started by looking at the components used to assist with the layout of our pages. We then looked at components that can help the user navigate around your application. Next, you learned how to collect user input using Material-UI form components. Finally, you learned how to style your Material-UI using styles and modifying themes.

In the next chapter, we'll look at ways to improve the efficiency of your component state updates using the latest functionality available in the latest version of React.

13
High-Performance State Updates

State represents the dynamic aspect of your React application. When state changes, your components react to those changes. Without state, you would have nothing more than a fancy HTML template language. Usually, the time required to perform a state update and have the changes rendered on the screen is barely noticeable if at all. However, there are times that complex state changes can lead to noticeable lag for your users.

In this chapter, you'll learn how to do the following:

- Batch your state changes together for minimal re-rendering.
- Prioritize state updates to render content that's critical for your user experience first.
- Develop strategies for performing asynchronous actions while batching and prioritizing state updates.

Technical requirements

For this chapter, you'll need your code editor (Visual Studio Code). The code we'll be following can be found here: `https://github.com/PacktPublishing/React-and-React-Native-4th-Edition/tree/main/Chapter13`.

You can open the terminal within Visual Studio Code and run `npm install` to make sure you're able to follow along with the examples as you read through the chapter.

Batching state updates

In this section, you'll learn about how React can batch state updates together in order to prevent unnecessary rendering when multiple state changes happen at the same time. In particular, we'll look at the changes introduced in React 18 that make automatic batching of state updates commonplace.

When your React component issues a state change, this causes the React internals to re-render the parts of your component that have changed visually as a result of this state update. For example, imagine you have a component with a name state that's rendered inside of a `<span>` element and you change the name state from `Adam` to `Ashley`. That's a straightforward change that results in a re-render that's too fast for the user to even notice. Unfortunately, state updates in web applications are rarely this straightforward. Instead, there might be dozens of state changes in 10 milliseconds. For example, the name state might follow changes like this:

1. `Adam`
2. `Ashley`
3. `Andrew`
4. `Ashley`
5. `Aaron`
6. `Adam`

Here, we have six changes that took place with the name state in a short amount of time. This means that React would have re-rendered the DOM six times, once for each value that was set as the name state. What's interesting to note about this scenario is the final state update – we're back where we started with `Adam`. This means that we just re-rendered the DOM five times for no reason. Now, imagine these wasted re-renders on a web application scale and how these types of state updates might cause problems for performance.

The answer to this problem is batching. This is how React takes several state updates that were made in our component code and treats them as a single state update. Rather than process every state update individually, while re-rendering the DOM between each update, the state changes are all merged, which results in one DOM re-render. In the aggregate, this reduces the amount of work that our web applications need to do by a lot.

In React 17, automatic batching of state updates only happened inside of event handler functions. For example, let's say you have a button with an `onClick()` handler that performs five state updates. React will batch all of these state updates together so that only one re-render is necessary. The problem arises when your event handlers make asynchronous calls, usually to fetch some data, and then make state updates when the asynchronous call finishes. These state changes are no longer automatically batched because they're not running directly inside of the event handler function. Instead, they're running in the callback code of the asynchronous operation, and React 17 will not batch these updates. This is a challenge because it's fairly common for our React components to fetch data asynchronously and perform state updates in response to events!

Let's turn our attention to some code now to see how React 18 addresses this batching problem that we've just outlined. For this example, we'll render a button that, when clicked, will perform 100 state updates. We'll use `setTimeout()` so that the updates are performed asynchronously, outside of the event handler function. The idea is to show the difference between how this code is handled by two different React versions. To do this, we can open up the React profiler in the browser dev tools and hit record before we press the button to execute our state changes. Here's what the code looks like:

```
import * as React from "react";

export default function BatchingUpdates() {
  let [value, setValue] = React.useState("loading...");

  function onStart() {
    setTimeout(() => {
      for (let i = 0; i < 100; i++) {
        setValue('value ${i + 1}');
      }
    }, 1);
  }

  return (
    <div>
      <p>
        Value: <em>{value}</em>
      </p>
      <button onClick={onStart}>Start</button>
    </div>
```

```
    );
}
```

By clicking the button that this component renders, we're calling the `onStart()` event handler function defined by our component. Then, our handler calls `setValue()` 100 times inside a loop. Ideally, we do not want to perform 100 re-renders because this will hurt the performance of our application, and it doesn't need to. Only the final call to `setValue()` matters here.

Let's take a look at the profile captured for this component using React 17:

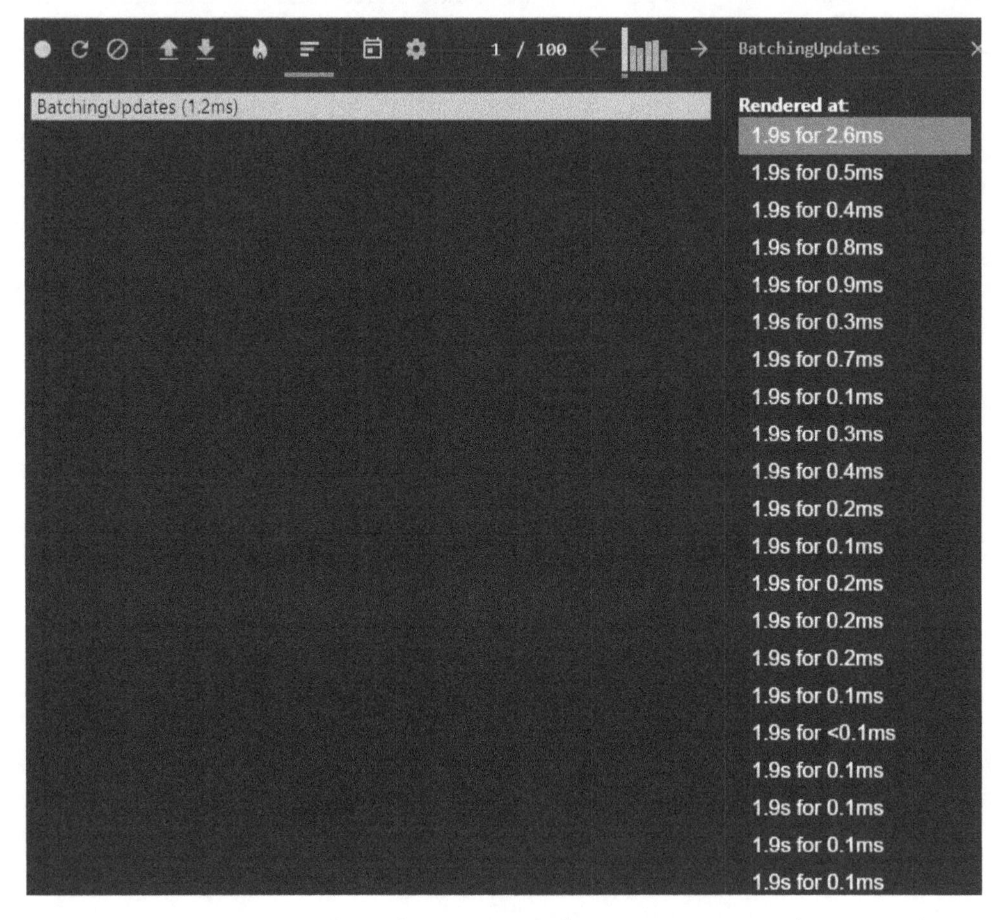

Figure 13.1 – Using React dev tools to view re-renders every time state updates are made

By pressing the button with our event handler attached to it, we're making 100 state update calls. Since this is done outside of the event handler function in `setTimeout()`, automatic batching doesn't happen. We can see this in the profile output of the `BactchingUpdates` component where there's a long list of renders. Most of these aren't necessary and contribute to the amount of work React needs to do in response to user interactions, hurting the overall performance of our application.

Let's capture a profile of the same component being rendered using React 18:

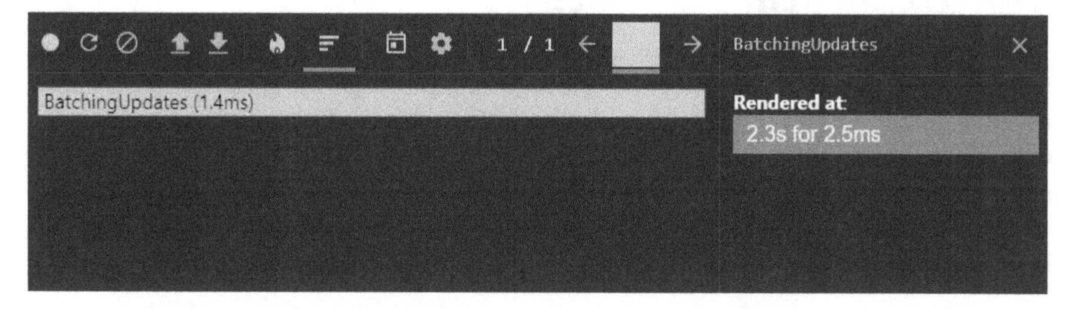

Figure 13.2 – React dev tools showing only one render with automatic batching enabled

Automatic batching is applied everywhere state updates are made, even in common asynchronous scenarios such as this one. As the profile shows, there's only one re-render when we click the button instead of 100. We didn't have to make any adjustments to our component code to make this happen either. However, there is one change that's required in order to make state updates batch automatically. Let's say you used `ReactDOM.render()` to render your root component, like so:

```
ReactDOM.render(
  <React.StrictMode>
    <App />
  </React.StrictMode>,
  document.getElementById("root")
);
```

Instead, you can use `ReactDOM.createRoot()` and render that:

```
ReactDOM.createRoot(document.getElementById("root")).render(
  <React.StrictMode>
    <App />
  </React.StrictMode>
);
```

By creating and rendering your root node this way, you can ensure that with React 18, you'll get batched state updates throughout your application. You no longer need to worry about manually optimizing state updates so that they take place at once – React does this for you now. However, sometimes you'll have state updates that are of higher priority than others. In cases like these, we need a way to tell React to prioritize certain state updates over others instead of batching everything together.

Prioritizing state updates

When something happens in our React application, we usually make several state updates so that the UI can reflect these changes. Typically, you can make these state changes without much thought about how the rendering performance is impacted. For example, let's say you have a long list of items that need to be rendered. This will probably have some impact on the UI – while the list is being rendered, the user probably won't be able to interact with certain page elements because the JavaScript engine is 100% utilized for a brief moment.

However, this can become an issue when expensive rendering disrupts the normal browser behavior that users expect. For example, if the user is typing in a textbox, they expect the character they just typed to show up immediately. But if your component is busy rendering a large item list, the textbox state cannot be updated right away. This is where the new React state update prioritization API comes in handy.

The `startTransition()` API is used to mark certain state updates as transitional, meaning that the updates are treated as a lower priority. If you think about a list of items either being rendered for the first time or being changed to another list of items, this is a transition that doesn't have to be immediate. On the other hand, state updates such as changing the value in a textbox should be as close to immediate as possible. By using `startTransition()`, you're telling React that any state updates within can wait if there are more important updates.

A good rule of thumb for `startTransition()` is to use it for the following:

- Anything that has the potential to perform a lot of rendering work
- Anything that doesn't require immediate feedback for the user in response to their interactions

Let's walk through an example that renders a large list of items in response to a user typing in a textbox to filter the list.

This component will render a textbox that the user can type in to filter the list of 25,000 items. I've chosen this number based on the performance of the laptop I'm using to write this code – you might want to tweak it up if there's no delay or down if it takes too long to render anything. When the page first loads, you should see a filter textbox that looks like this:

Figure 13.3 – The filter box before the user types anything

When you start typing in the filter textbox, the filtered items will render underneath it. It might take a second or 2, since there are so many items to render:

1

- Item 1
- Item 10
- Item 11
- Item 12
- Item 13
- Item 14
- Item 15
- Item 16

Figure 13.4 – Filtered items underneath the filter input when the user starts typing

Now, let's walk through the code, starting with a large array of items:

```
let unfilteredItems = new Array(25000)
  .fill(null)
  .map((v, i) => ({ id: i, name: 'Item ${i}' }));
```

The size of the array is specified in the array constructor, and then it's filled with numbered string values that we can filter by.

Next, let's look at the state used by this component:

```
let [filter, setFilter] = React.useState("");
let [items, setItems] = React.useState([]);
```

The `filter` state represents the value of the filter textbox and defaults to an empty string. The `items` state represents the filtered items from our `unfilteredItems` array. This array is populated when the user types in the filter textbox.

Next, let's look at the markup rendered by this component:

```
<div>
  <div>
    <input
      type="text"
      placeholder="Filter"
      value={filter}
      onChange={onChange}
    />
  </div>
  <div>
    <ul>
      {items.map(item => (
        <li key={item.id}>{item.name}</li>
      ))}
    </ul>
  </div>
</div>
```

The filter textbox is rendered by an `<input>` element while the filtered results are rendered as a list by iterating over the `items` array.

Finally, let's look at the event handler function that's fired when the user types in the filter textbox:

```
function onChange(e) {
  setFilter(e.target.value);
  setItems(
    e.target.value === ""
      ? []
      : unfilteredItems.filter(
        item => item.name.includes(e.target.value)
      )
```

```
  );
}
```

The `onChange()` function is called when the user types in the filter textbox and sets two state values. First, it uses `setFilter()` to set the value of the filter textbox. Then, it calls `setItems()` to set the filtered items to render unless the filter text is empty, in which case, we render nothing. When interacting with this example, you might notice a problem with the responsiveness of the textbox when typing in it. This is because, in this function, we're setting not only the textbox value but also the filtered items. This means that before the text value can be rendered, we have to wait for thousands of items to be rendered.

Even though these are two separate state updates (`setFilter()` and `setItems()`) they're batched and treated as a single state update. Likewise, when the rendering starts, React is making all the changes at once, which means that the CPU won't let the user interact with the textbox because it's fully utilized, rendering the long list of filter results. Ideally, we want to prioritize the textbox state update while letting the items render afterward. To put it another way, we want to deprioritize the item rendering, since it's expensive and the user isn't interacting with it directly.

This is where the `startTransition()` API comes in. Any state updates that take place within the function that's passed to `startTransition()` will be treated with lower priority than any state updates that happen outside of it. In our filtering example, we can fix the textbox responsiveness issue by moving the `setItems()` state change inside of `startTransition()`.

Here's what our new `onChange()` event handler looks like:

```
function onChange(e) {
  setFilter(e.target.value);
  React.startTransition(() => {
    setItems(
      e.target.value === ""
        ? []
        : unfilteredItems.filter(
          item => item.name.includes(e.target.value)
        )
    );
  });
}
```

Note that we didn't have to make any changes to how the `items` state is updated – the same code is moved to a function that's passed to `startTransition()`. This tells React to only execute this state change after any other state changes are complete. In our case, this allows the textbox to update and render before the `setItems()` state change runs. If you run the example now, you'll see that the responsiveness of the textbox is no longer affected by how long it takes to render a long list of items.

Before this new API was introduced, you could achieve state update prioritizations via workarounds with `setTimeout()`. The main disadvantage of this approach is that the internal React scheduler knows nothing about your state updates and their priorities. For example, by using `startTransitiion()`, React can cancel the update entirely if the state changes again before completion or if the component is unmounted.

In real applications, it isn't simply a matter of prioritizing which state updates should run first. Rather, it's a combination of fetching data asynchronously while making sure that priorities are taken into account. In the final section of this chapter, we'll tie all of this together.

Handling asynchronous state updates

In this final section of the chapter, we'll look at the common scenario of fetching data asynchronously and setting render priorities. The key scenario that we want to address is making sure that users aren't interrupted from typing or any other interaction that requires immediate feedback. This requires both proper prioritization and handling asynchronous responses from the server. Let's start by looking at the React APIs that can potentially help with this scenario.

The `startTransition()` API can be used as a Hook. When we do this, we also get a Boolean value that we can check to see whether the transition is still pending. This is useful for showing the user that things are loading. Let's modify the example from the previous section to use an asynchronous data-fetching function for our items. We'll also use the `useTransition()` Hook and add loading behavior to the output of our component:

```
import * as React from "react";

let unfilteredItems = new Array(25000)
  .fill(null)
  .map((v, i) => ({ id: i, name: 'Item ${i}' }));

function filterItems(filter) {
```

```
  return new Promise(resolve => {
    setTimeout(() => {
      resolve(unfilteredItems.filter(
        item => item.name.includes(filter)));
    }, 1000);
  });
}

export default function AsyncUpdates() {
  let [isPending, startTransition] = React.useTransition();
  let [filter, setFilter] = React.useState("");
  let [items, setItems] = React.useState([]);

  async function onChange(e) {
    setFilter(e.target.value);

    startTransition(async () => {
      setItems(e.target.value === ""
        ? []
        : await filterItems(e.target.value)
      );
    });
  }

  return (...)

}
```

What this example shows is that once you start typing in the filter textbox, this will trigger the `onChange()` handler, which will call the `filterItems()` function. Since we're using the `useTransition()` Hook, we have an `isPending` value that we can use to show the user that something is happening in the background:

```
<div>
  <div>
    <input
      type="text"
```

```
        placeholder="Filter"
        value={filter}
        onChange={onChange}
      />
    </div>
    <div>
      {isPending && <em>loading...</em>}
      <ul>
        {items.map(item => (
          <li key={item.id}>{item.name}</li>
        ))}
      </ul>
    </div>
  </div>
```

Here's what the user will see when `isPending` is `true`:

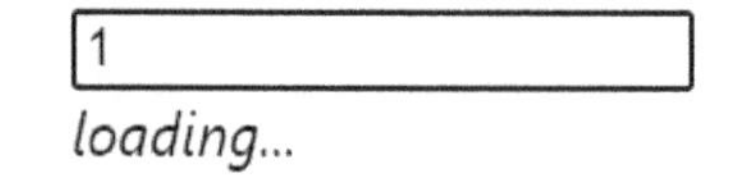

Figure 13.5 – A loading indicator while a state transition is pending

However, there's a slight problem with our approach. You might have noticed the loading message flash briefly when typing into the textbox. But then, you probably had a longer period where the items still weren't visible and the loading message disappeared. What's happening here? Well, the `isPending` value that comes from the `useTransition()` Hook can be misleading. We've designed our component in such a way that `isPending` will be `true` in the following situations:

- If the `filterItems()` function is still fetching our data
- If the `setItems()` state update is still performing the expensive render with lots of items

Unfortunately, this isn't how `isPending` works. The only time this value is `true` is before the function we pass to `startTransition()` is run. This is why you'll see the loading indicator flash briefly instead of being displayed throughout the data-fetching operation and the rendering operation. Remember, React schedules state updates internally, and by using `startTransition()`, we've scheduled `setItems()` to run after other state updates.

Another way to think about `isPending` is that it's `true` while high-priority updates are still running. We can call it `highPriorityUpdatesPending` to avoid confusion. That said, the uses of this value are narrow, but they do happen from time to time. For our more common case of fetching data and performing an expensive render, we need to think of another solution. Let's walk through our code and refactor it in such a way that the loading indicator is displayed while the fetch is happening and while the higher-priority updates happen. First, let's introduce a new `loading` state that defaults to `false`:

```
let [loading, setLoading] = React.useState(false);
let [filter, setFilter] = React.useState("");
let [items, setItems] = React.useState([]);
```

Now, inside of our `onChange()` handler, we can set this to `true`. Inside of our transition that runs after the data fetch completes, we set it back to `false`:

```
async function onChange(e) {
  setLoading(true);
  setFilter(e.target.value);

  React.startTransition(async () => {
    setItems(e.target.value === ""
      ? []
      : await filterItems(e.target.value)
    );
    setLoading(false);
  });
}
```

Now that we're keeping track of the `loading` state, we know exactly when all the heavy lifting is done and can hide the loading indicator. The final change is to base the indicator display on `loading` instead of `isPending`:

```
<div>
  {loading && <em>loading...</em>}
  <ul>
    {items.map(item => (
      <li key={item.id}>{item.name}</li>
    ))}
```

```
    </ul>
  </div>
```

When you run the example with these changes, the results should be a lot more predictable. The `setLoading()` and `setFilter()` state updates are high-priority and execute immediately. The call to fetch data using `filterItems()` isn't made until the high-priority state updates are completed. Only after we have the data do we hide the loading indicator.

Summary

This chapter introduced you to the new APIs available in React 18 that help you achieve high-performance state updates. We started with a look at the changes to automatic state update batching in React 18 and how to best take advantage of them. We then explored the new `startTransition()` API and how it can be used to mark certain state updates as having a lower priority than those that require immediate feedback for user interactions. Finally, we looked at how state update prioritization can be combined with asynchronous data fetching.

In the next chapter, we'll go over what makes React Native a good choice for native application development.

Part 2 – React Native

In this part, we look at building mobile apps with the React Native library. We'll explore the basic API and some common approaches to help you develop solid and performant applications.

This part contains the following chapters:

- *Chapter 14, Why React Native?*
- *Chapter 15, React Native under the Hood*
- *Chapter 16, Kick-Starting React Native Projects*
- *Chapter 17, Building Responsive Layouts with Flexbox*
- *Chapter 18, Navigating between Screens*
- *Chapter 19, Rendering Item Lists*
- *Chapter 20, Showing Progress*
- *Chapter 21, Geolocation and Maps*
- *Chapter 22, Collecting User Input*
- *Chapter 23, Displaying Modal Screens*
- *Chapter 24, Responding to User Gestures*
- *Chapter 25,* Using *Animations*
- *Chapter 26, Controlling Image Display*
- *Chapter 27, Going Offline*
- *Chapter 28, Selecting Native UI Components Using NativeBase*

14
Why React Native?

Facebook created **React Native** (**RN**) to build its mobile applications. It started as a hackathon project in the summer of 2013 inside Facebook and became open source for everyone in 2015. The motivation to release it was because React for the web was so successful. They thought that if React was such a good tool for **user interface** (**UI**) development, and you wanted a native application, why not just make React work with mobile OS UI elements!

Therefore, in the same year, Facebook divided **React** into two independent libraries, **React** and **ReactDOM**, and since then, React has had to work only with interfaces and not care about where these elements will be rendered. The rendering part for the web was taken by ReactDOM, and for mobile platforms by RN.

In this chapter, you'll learn about the motivations for using RN to build native mobile web applications. Here are the topics that we'll cover in this chapter:

- What is RN?
- React and JSX are familiar
- The mobile browser experience
- Android and iOS – different yet the same
- The case for mobile web apps

Technical requirements

There aren't any technical requirements for this chapter, since it is a brief conceptual introduction to RN.

What is RN?

RN is a JavaScript-based mobile app framework that allows you to create natively rendered mobile apps for iOS and Android. Frameworks allow you to create an application for multiple platforms using the same code base.

Earlier in this book, I introduced the notion of a render target—the thing that React components render to. The render target is abstract as far as the React programmer is concerned. For example, in React, the render target can be a string, or it could be the **Document Object Model** (**DOM**). Therefore, your components never directly interface with the render target, because you can never make assumptions about where the rendering is taking place.

A mobile platform has UI widget libraries that developers can leverage to build apps for that platform. On Android, developers implement Java apps, while on iOS, developers implement Swift apps. If you want a functional mobile app, you're going to have to pick one. However, you'll need to learn both languages, as supporting only one of two major platforms isn't realistic for success.

For React developers, this isn't a problem. The same React components that you build work all over the place, even on mobile browsers! Having to learn two more programming languages to build and ship a mobile application is costly and time-intensive. The solution to this is to introduce a new React platform that supports a new render Replace with target-native mobile UI widgets.

RN uses a technique that makes asynchronous calls to the underlying mobile OS, which calls the native widget APIs. There's a JavaScript engine, and the React API is mostly the same as React for the web. The difference is with the target; instead of a DOM, there are asynchronous API calls. The concept is visualized here:

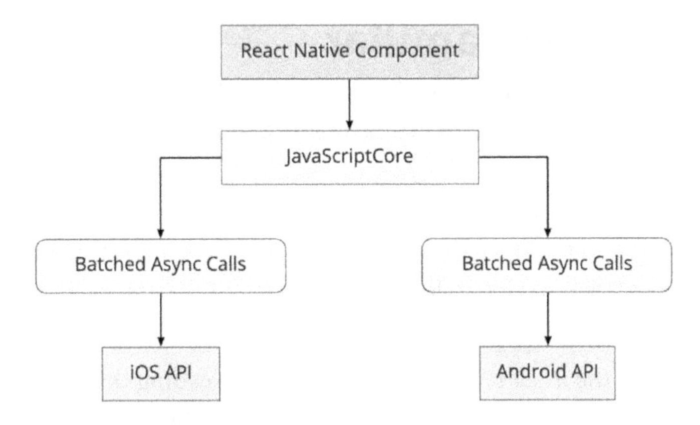

This oversimplifies everything that's happening under the hood, but the basic ideas are as follows:

- The same React library that's used on the web is used by RN and runs in **JavaScriptCore**.
- Messages that are sent to native platform APIs are asynchronous and batched for performance purposes.
- RN ships with components implemented for mobile platforms, instead of components that are HTML elements.
- RN represents a way to render components via iOS and Android APIs. It can be replaced using the same concept with tvOS, Android TV, Windows, macOS, and even Web again. This is possible by using forks and add-ons for RN. In this part of the book, we will learn how to write mobile apps for iOS and Android. More information about other possible platforms can be found here: `https://reactnative.dev/docs/out-of-tree-platforms`.

> **Important Note**
>
> Much more on the history and mechanics of RN can be found at `https://engineering.fb.com/2015/03/26/android/react-native-bringing-modern-web-techniques-to-mobile/`.

In the next chapter, we'll take a closer look at each part of the RN architecture. Now that you know what RN is, it's time to look at what attracts React developers to it.

React and JSX are familiar

Implementing a new render target for React is not straightforward. It's essentially the same thing as inventing a new DOM that runs on iOS and Android. So, why go through all the trouble?

First, there's a huge demand for mobile apps. The reason is that the mobile web browser user experience isn't as good as the native app experience. Second, JSX is a fantastic tool for building UIs. Rather than having to learn new technology, it's much easier to use what you know.

It's the latter point that's the most relevant to you. If you're reading this book, you're probably interested in using React for both web applications and native mobile applications. I can't put into words how valuable React is from a development-resource perspective. Instead of having a team that does web UIs, a team that does iOS, a team that does Android, and so on, there's just the UI team that understands React.

In the following section, you'll learn about the challenges of delivering good user experiences on mobile web browsers.

The mobile browser experience

Mobile browsers lack many capabilities of mobile applications. This is because browsers cannot replicate the same native platform widgets as HTML elements. You can try to do this, but it's often better to just use the native widget rather than try to replicate it. This is partly because this requires less maintenance effort on your part, and partly because using widgets that are native to the platform means that they're consistent with the rest of the platform. For example, if a date picker in your application looks different from all the date pickers the user interacts with on their phone, this isn't a good thing. Familiarity is key and using native platform widgets makes familiarity possible.

User interactions on mobile devices are fundamentally different from the interactions that you typically design for the web. Web applications assume the presence of a mouse, for example, and that the click event on a button is just one phase. However, things become more complicated when the user uses their fingers to interact with the screen. Mobile platforms have what's called a **gesture system** to deal with this. RN is a much better candidate for handling gestures than React for the web because it handles these types of things that you don't have to think about much in a web app.

As the mobile platform is updated, you want the components of your app to stay updated too. This isn't a problem with RN because they're using actual components from the platform. Once again, consistency and familiarity are important for a good user experience. So, when the buttons in your app look and behave in the same way as the buttons in every other app on the device, your app feels like part of the device.

Now that you understand what makes developing UIs for mobile browsers difficult, it's time to look at how RN can bridge the gap between the different native platforms.

Android and iOS - different yet the same

When I first heard about RN, I automatically thought that it would be some cross-platform solution that lets you write a single React application that will run natively on any device. Do yourself a favor and get out of this mindset before you start working with RN. iOS and Android are different on many fundamental levels. Even their user experience philosophies are different, so trying to write a single app that runs on both platforms is categorically misguided.

Besides, this is not the goal of RN. The goal is to *learn once and write anywhere*, not write once, run anywhere. In some cases, you'll want your app to take advantage of an iOS-specific widget or an Android-specific widget. This provides a better user experience for that platform and should trump the portability of a component library.

There are several areas that overlap between iOS and Android where the differences are trivial. The two widgets aim to accomplish the same thing for the user, in roughly the same way. In these cases, RN will handle the difference for you and provide a unified component. In the next section, we'll look at the case where mobile web apps that run in the browser might be a better fit for your users.

The case for mobile web apps

Not every one of your users is going to be willing to install an app, especially if you don't yet have a high download count and rating. The barrier to entry is much lower with web applications – the user only needs a browser.

Despite not being able to replicate everything that native platform UIs have to offer, you can still implement awesome things in a mobile web UI. Maybe having a good web UI is the first step toward getting those download counts and ratings up for your mobile app.

Ideally, you should aim for the following:

- Standard web (laptop/desktop browsers)
- Mobile web (phone/tablet browsers)
- Mobile apps (phone-/tablet-native platform)

Putting an equal amount of effort into all three of these spaces might not make much sense, as your users probably favor one area over another. Once you know, for example, that there's a high demand for your mobile app compared to the web versions, that's when you allocate more effort there.

Summary

In this chapter, you learned that RN is an effort by Facebook to reuse React to create native mobile applications. React and JSX are good at declaring UI components, and since there's now a huge demand for mobile applications, it makes sense to use what you already know for the web.

The reason there's such a demand for mobile applications over mobile browsers is that they just feel better. Web applications lack the ability to handle mobile gestures the same way apps can, and they generally don't feel like part of the mobile experience from a look and feel perspective.

RN isn't trying to implement a component library that lets you build a single React app that runs on any mobile platform. iOS and Android are fundamentally different in many important ways. Where there's overlap, RN does try to implement common components. Will you do away with mobile web apps now that we can build natively using React? This will probably never happen because the user can only install so many apps.

Now that you know what RN is and what its strengths are, you'll learn how RN works under the hood.

Further reading

You can find more information on RN at `https://reactnative.dev/`.

15
React Native under the Hood

The previous chapter briefly touched on what React Native is and the differences that users experience between the Native UI and mobile browsers.

In this chapter, we will dig deeper into React Native as well as become well versed on how it performs on mobile devices and what we should attain before commencing any efforts with this framework. We will also look at what options we can execute for the native functionality of JavaScript and what restrictions we will come up against.

So, in this chapter, we will cover the following topics:

- Exploring React Native architecture
- Explaining JavaScript and Native modules
- Exploring React Native components and APIs

> **Important Note**
> Meta is the company formerly known as Facebook before 2021.

Technical requirements

This chapter doesn't have any technical requirements since it is an introduction to React Native.

Exploring React Native architecture

Before understanding how React Native works, let's revise some history points about React architecture and the differences between web and native mobile apps.

The state of web and mobile apps in the past

Meta released React in 2013 as a monolith tool for creating apps using a component approach and a virtual DOM. It gave us the opportunity to develop web applications without thinking about browser processes, such as how it parses JS code, and creates the DOM, and layers and rendering. We just had to create interfaces using state and props for data and CSS for styling, fetch data from the backend, save it in local storage, and so on. React, together with browsers, allowed us to create a performance application in less time. At that time, the architecture of React looked like this:

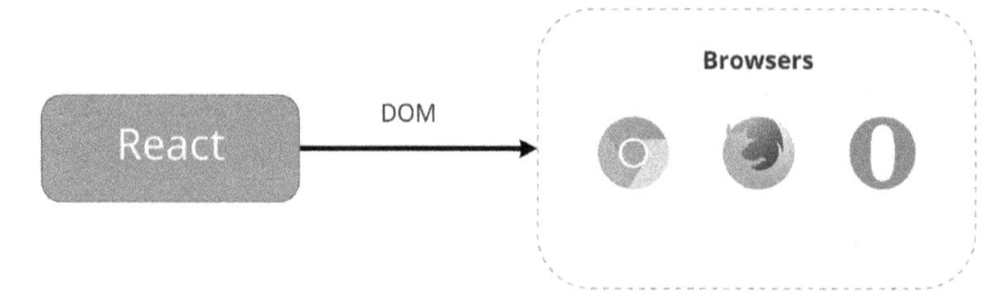

Figure 15.1 – React architecture in 2013

The new declarative approach to developing interfaces became more favorable because of the fast development and the low threshold for novices. Additionally, if your backend is built with Node.js, you can benefit from the ease of support and development of the entire project with just one developer.

At the same time, mobile apps require more complex techniques to create the apps. For Android and iOS apps, companies should manage three different teams with unparalleled experience to support three major ecosystems:

- Web developers should know HTML, CSS, JS, and React.
- Java or Kotlin SDK experience is required for Android developers.
- The iOS developer should be familiar with Objective-C or Swift and CocoaPods.

Every step of developing an application from prototyping to release requires unique skills. Web and mobile app development before cross-platform solutions looked like this:

Figure 15.2 – The state of web and mobile apps

Even if a corporation carries out a basic application, it can be faced with some major issues:

- Each of these teams implements the same business logic.
- There is no alternative to sharing code between teams.
- It is not conceivable to share resources between teams (Android developers can't write code for iOS applications and vice versa).

As a result of these significant issues, we likewise have complications with having more testing resources, since there are more places to create bugs. The speed of development is also diverse because mobile apps take more time to deliver the same features. This all accumulates into a large, costly problem for the companies involved. Many of them came up with ideas on how to write a single code base or reuse a current one that can be used in multiple ecosystems. The simplest method would be to wrap a web app to mobiles using a browser, but this has limitations in handling touch and gestures (we explored this more in the previous chapter).

Following these issues, Meta started investing resources in developing a cross-platform framework and released the React Native library in 2015. Also, they divided React into two separate libraries. For rendering our app in the browser, we should now use the ReactDOM library.

In *Figure 15.3*, we can see how React works together with ReactDOM and React Native to render our apps:

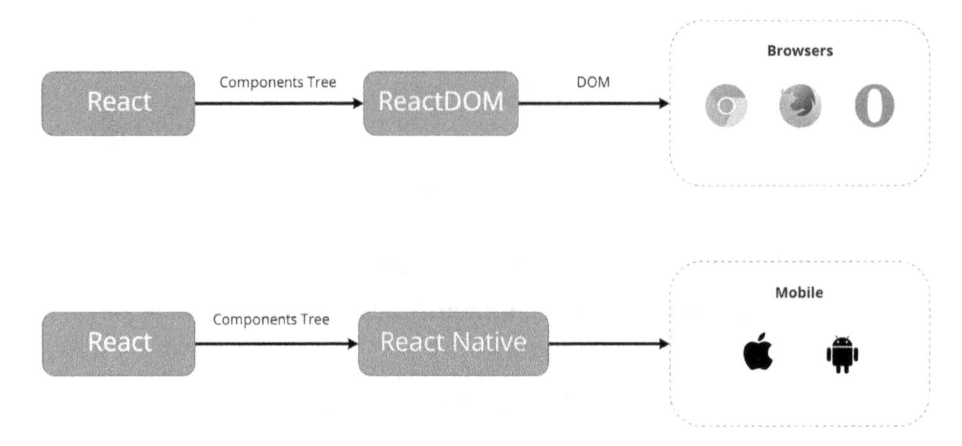

Figure 15.3 – React and React Native flow

Now, React only works for managing components tree. This approach incapsulates any rendering APIs and hides a lot of platform-specific methods from us. We can concentrate solely on developing interfaces and cease speculating about how they would be rendered. That's why React is frequently claimed as a renderer-agnostic library. And for web apps, we use ReactDOM, which forms elements and applies them right to the browser DOM. For mobile apps, React Native renders our interface directly on the mobile screen.

But how does React Native replace the whole browser API and allow us to write familiar code and run it on mobiles?

React Native current architecture

The React Native library allows you to create native applications with React and JS by utilizing **native** building blocks. For instance, the `<Image/>` component represents two other native components, `ImageView` on Android and `UIImageView` on iOS. This is viable because of the architecture of React Native, which includes two dedicated layers, represented by JS and Native threads:

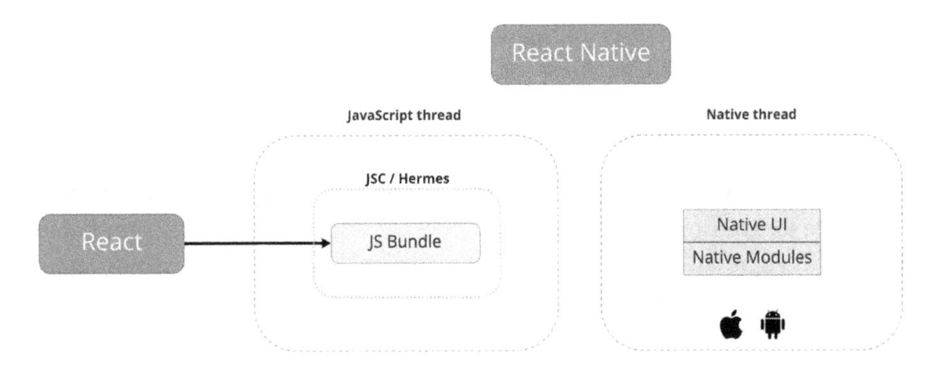

Figure 15.4 – React Native threads

In the next sections, we will explore each thread and see how they can communicate, ensuring that JS is integrated into the native code.

JS part of React Native

As the browser executes JS through a JS virtual machine such as V8, SpiderMonkey, and others, React Native also contains a JS virtual machine. There, our JS code is executed, API calls are made, touch events are processed, and many other processes occur.

Initially, React Native only supported Apple's JSCore virtual machine. With iOS devices, this virtual machine is built-in and available out of the box. In the case of Android, JSCore is bundled with React Native. This increases the size of the app. Therefore, the Hello World application of React Native would consume approximately 3 to 4 MB on Android. From the 0.60 version, React Native started using the new Hermes virtual machine, and from 0.64, for iOS as well.

The Hermes virtual machine introduced a lot of improvements for both platforms:

- Improvement of the app's start up time
- Reduced size of the downloaded app
- Decreased memory usage
- Built-in Proxy support, enabling the use of `react-native-firebase` and `mobx`

> **Note**
> More information about Hermes can be found here: `https://reactnative.dev/docs/hermes`.

JS in React Native, as in browsers, is implemented in a single thread. That thread is responsible for executing JS. The business logic we are writing is carried out in this thread. It means all our common code, such as components, state, hooks, and REST API calls, will be handled in the JS part of the app.

Our entire application structure is packaged into a single file using the Metro bundler. It is also responsible for transpiling JSX code into JS. If we want to use TypeScript, Babel can support it. It works right out of the box, so there's no need to configure anything. In future chapters, we will learn how to start a ready-to-work project.

The Native part

Here is where native code is executed. React Native implements this part in native code for each platform, Java for Android and Objective-C for iOS. The Native layer is mainly composed of Native modules that communicate with the Android or iOS SDK and are supposed to provide native functionality to our apps using a unified API. If we want to display an alert dialog, for instance, the Native layer presents a unified API for both platforms, which we will call from the JS thread using the single API.

This thread interacts with the JS thread when you need to update the interface or call the native functions. There are two parts to this thread:

- The first, Native UI, is responsible for using native interface shaping tools.
- The second is Native Modules, which allow applications to access specific capabilities of the platform on which they run.

Communication between threads

As previously mentioned, each React Native layer implements a unique API for every native and UI feature in the application. The communication between layers is accomplished through the bridge. The module is written in C ++ and is based on an asynchronous queue. When the bridge receives data from one of the parties, it serializes it, converts it to a JSON string, and passes it through the queue. After arriving at its destination, the data is deserialized.

As shown in the alert example, the native part accepts the call from JS and displays the dialog. In reality, the JS method, upon being invoked, sends a message to the bridge, and upon receiving this message, the native part executes the instruction. Native messages may also be forwarded to the JS layer. On clicking the button, for example, the Native layer sends a message to the JS one with an `onClick` event. It can be imagined as follows:

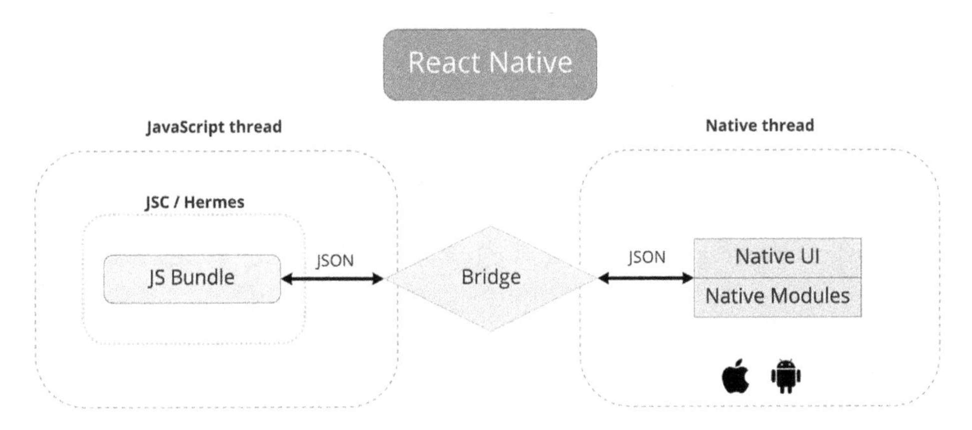

Figure 15.5 – The bridge

JS and the Native part of this architecture, together with the bridge, resemble the server and client sides of web applications, where they communicate through the REST APIs. It does not matter to us in what language or how the Native part is implemented, since the code in JS is isolated. We simply send messages and receive responses from the bridge. This is both a significant advantage and a great disadvantage – first, it allows us to implement cross-platform apps with one code base, but it can be a bottleneck in our app when we have a lot of business logic in it. All events and actions in the application rely on asynchronous JSON bridged messages. Each party sends these messages, expecting (which is not guaranteed) that sometime in the future, a response will be received from these messages. With such a data exchange scheme, there is a risk of overloading the communication channel.

Here is an example commonly used to illustrate how such a communication scheme can cause performance problems for an application. Suppose a user of an application is scrolling through a huge list. When the `onScroll` event occurs in the native environment, information is passed asynchronously to the JS environment. But native mechanisms do not wait until the JS part of the application does its job and reports to them about it. Because of this, there is a delay in the appearance of empty space in the list before displaying its contents. We can avoid a lot of usual problems using special approaches, such as using paginated `FlatList` on limitless lists. We will look at the main tricks in future chapters, but it is important to remember the limitations of the current architecture.

Styling

As we already understand the concept of cross-platform, we can assume that each platform has its own technologies for creating and styling interfaces. In order to unify this, React Native has a CSS-in-JS syntax for styling the app. Using Flexbox, components are able to specify the layout of their children. This ensures a consistent layout across different screen sizes. It is usually similar to how CSS works on the web, except the names are written in camel case, such as `backgroundColor` rather than `background-color`.

In JS, it is a plain object with style properties, and in native code, it is a separate thread called Shadow. It recalculates the layout of the application using the **Yoga** engine, which is developed by Meta. In this thread, the calculations related to the formation of the application interface are performed. The results of these calculations are sent to the Native UI thread responsible for displaying the interface.

With all the parts coming together, the final architecture of React Native is illustrated in this figure:

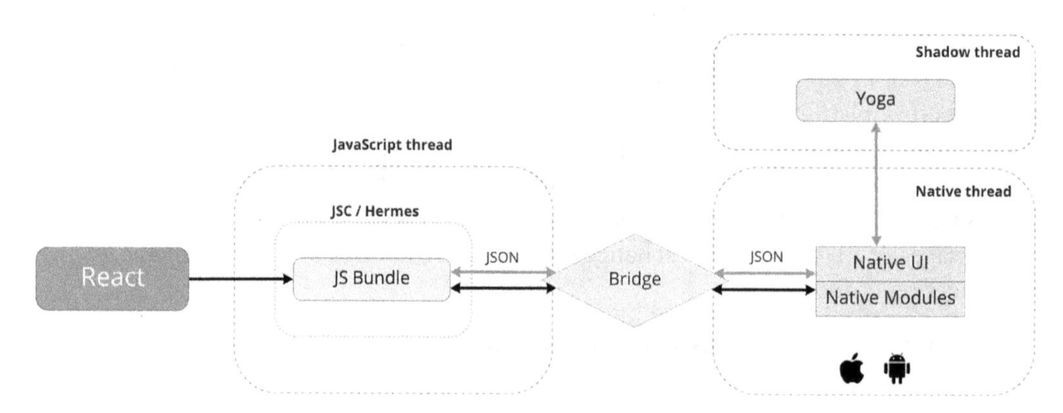

Figure 15.6 – The current React Native architecture

The current architecture of React Native addresses major business problems – it is feasible to develop web and mobile applications within the same team, it is possible to reuse a large amount of business logic code, and even developers with no previous experience in mobile development can easily use React Native.

However, the current architecture is not ideal. Over the past few years, the React Native team has been working on a bridge bottleneck solution. The new architecture is designed to address this issue.

React Native future architecture

A series of significant improvements have been introduced with React Native that will streamline the development process and make it more convenient for everyone.

React Native re-architecture will gradually deprecate the bridge and replace it with a new component called the **JS Interface (JSI)**. In addition, this element will enable new Fabric and TurboModules.

The use of the JSI opens up many possibilities for improvement. In *Figure 15.7*, you can see the major updates to the React Native architecture:

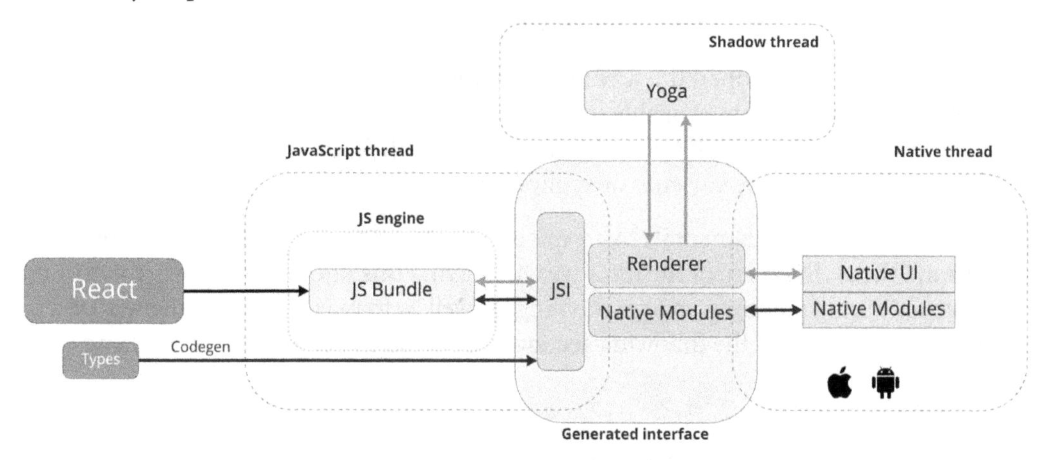

Figure 15.7 – The new React Native architecture

The first change is that the JS bundle is no longer dependent on a JavaScriptCore virtual machine. It is actually part of the current architecture because now we can enable the new Hermes JS engine on both platforms. In other words, the JSC engine can now easily be replaced with something else, quite possibly with better performance.

The second improvement is what lies at the heart of the new React Native architecture. The JSI allows JS to call native methods and functions directly. This was made possible by the **HostObject** C++ object, which stores references to native methods and properties. HostObject in JS binds native methods and props to a global object, so direct calls to JS functions will invoke Java or Objective-C APIs.

Another benefit of the new React Native is the ability to fully control native modules called **TurboModules**. Rather than starting them all at once, the application will only use them when they are needed.

Fabric is the new UI manager, called **Renderer** in *Figure 15.7*, which is expected to transform the rendering layer by eliminating the need for bridges. It is now possible to create Shadow Tree directly in C++, which brings speed and reduces the number of steps to render a particular element.

In order to ensure smooth communication between React Native and Native parts, Meta is currently working on a tool called CodeGen. It is expected to automate the compatibility of strongly typed native code and dynamic typed JS to make them synchronized. To achieve this, the React Native team has developed a Codegen tool that will define the interface and types needed by TurboModules and Fabric. With this upgrade, there will be no need to duplicate the code for both threads, thereby enabling smooth synchronization.

The new architecture could open the way for the development of new designs that are capable of things that were not available in old RN applications. The fact is that now we have at our disposal the power of C++. This means that with React Native, it will now be possible to create many more varieties of applications than before.

Here, we discussed the fundamentals that explain how React Native works. It is important to understand the architecture of the tools we use. Having this knowledge allows you to avoid mistakes during planning and prototyping, as well as to maximize the potential of your future applications. In the following section, we will briefly explore how to extend React Native with modules.

Explaining JS and Native modules

React Native does not cover all native capabilities out of the box. It only provides the most common features that you will need in the basic application. Also, the Meta team itself has recently moved some functions into its own modules in an effort to reduce the size of the overall application. For example, **AsyncStorage** for storing data on the device was moved into a separate package and must be installed if you plan to use it.

However, React Native is an extendable framework. We can add our own native modules and expose the JS API using the same bridge or JSI. Our focus in this book will not be on developing native modules, since we need prior experience with Objective-C or Java. And it is not necessary, since the React community has created an enormous number of ready-to-use modules for all cases. We will learn how to install native packages in subsequent chapters.

The following are a few of the most popular native modules, without which most projects couldn't prosper.

React Navigation

React Navigation is one of the best React Native navigation libraries for creating navigation menus and screens for your app. It's a good tool for beginners because it's stable, fast, and less buggy. The documentation is really good, and it provides examples for all use cases.

We'll learn more about React Navigation in *Chapter 16, Kick-Starting React Native Projects*.

UI component libraries

The **UI component libraries** enable you to quickly assemble the application layout without wasting time designing and coding atomic elements. In addition, such libraries are often more stable and consistent, which leads to better results both in terms of UI and UX.

These are some of the most popular libraries (we will explore a few of them in greater detail in future chapters):

- NativeBase – this is a component library that enables developers to build universal design systems. It is built on top of React Native, allowing you to develop apps for Android, iOS, and the web.
- React Native Element – this provides an all-in-one UI kit for creating apps in React Native.
- UI Kitten – this is a React Native implementation of the Eva Design System. The framework contains a set of general-purpose UI components styled in a similar way.
- React-native-paper – this is a collection of customizable and production-ready components for React Native, following Google's Material Design guidelines.

Splash screen

Adding a **splash screen** to your mobile app can be a tedious task, since this screen should appear before the JS thread begins. The `react-native-bootsplash` package allows you to create a fancy splash screen from the command line. The package will do all the work for you if you provide it with an image and a background color.

Icons

Icons are an integral part of the visualization of interfaces. Different approaches are used to display icons and other vector graphics on each platform. React Native unifies this for us but only with additional libraries such as `react-native-vector-icons`. Using `react-native-svg`, you can also render SVG into a React Native app.

Handling errors

Usually, when we develop a web application, we are able to handle errors without any difficulty, since they do not reach beyond the scope of JS. As a result, we have more control and stability in the event of critical bugs because if the application does not start at all, we can easily see the reason, and open the logs in the DevTools.

There are even more complications with React Native applications, since we have a Native component, in addition to the JS of the environment, which can also cause errors in application execution. Therefore, when an error occurs, our application will close immediately. It will be hard for us to figure out why.

`React-native-exception-handler` provides a simple technique for handling native errors and JS errors and providing feedback. To make it work, you need to install and link the module. Then, register your global handler for JS and native exceptions, as follows:

```
import { setJSExceptionHandler, setNativeExceptionHandler }
  from "react-native-exception-handler";

setJSExceptionHandler((error, isFatal) => {
  // …
});

const exceptionhandler = (exceptionString) => {
  // your exception handler code here
};

setNativeExceptionHandler(
  exceptionhandler,
  forceAppQuit,
  executeDefaultHandler
);
```

The `setJSExceptionHandler` and `setNativeExceptionHandler` methods are custom global error handlers. If a crash occurs, you can show an error message, use Google Analytics to track it, or use a custom API to inform the development team.

Push notifications

We live in a world where notifications are integral. We open dozens of apps every day just because we receive notifications from them.

Push notifications are often connected to a gateway provider that sends messages to users' devices. The following libraries can be used to add push notifications to your application:

- `react-native-onesignal` – a OneSignal provider for push notifications, email, and SMS
- `react-native-firebase` – Google Firebase
- `@aws-amplify/pushnotification` – AWS Amplify

Over the air updates

As part of the normal application update, when you build a new version and upload it to the app store, you can replace the JS package in **Over the Air** (**OTA**). As the bundle contains only one file, updating it is not complicated. You can update your application as often as you like without waiting for Apple or Google to verify your application. That is the real power of React Native.

We can use it due to the CodePush service made by Microsoft. You can find more information about CodePush here: `https://docs.microsoft.com/en-gb/appcenter/distribution/codepush/`.

JS libraries

As for the JS (non-native) modules, we have almost no restrictions, except for libraries that use unsupported APIs, such as the DOM and Node.js. We can use any packages written in JS – Moment, Lodash, Axios, Redux, MobX, and a thousand others.

We have barely scratched the surface of the possibilities for extending the application with various modules in this section. Because React Native has thousands of libraries, it makes little sense to go through them all. In order to find the required package you need, there is a project called React Native Directory that has collected and rated a huge list of packages. The project can be found here: `https://reactnative.directory/`.

We now know how React Native is organized internally and how we can expand its functionality. Our next step is to examine what API and components this framework offers.

Exploring React Native components and APIs

The main modules and components will be discussed in detail in each new chapter, but for now, let's familiarize ourselves with them. A number of core components are available in the React Native framework for use in the app.

Almost all apps use at least one of these components. These are the fundamental building blocks of React Native apps:

- **View**: The main brick of any app. This is the equivalent of `<div>`, and on mobiles, it is represented as `UIView` and `android.view`. Any `<View/>` component can nest inside another `<View/>` component and can have zero or many children of any type.
- **Text**: This is a React component for displaying text. As with the View, `<Text/>` supports nesting, styling, and touch handling.
- **Image**: This displays images from a variety of sources, such as network images, static resources, temporary local images, and images from the camera roll.
- **TextInput**: This allows users to input text using a keyboard. Props enable a variety of features that can be configured, including auto-correction, auto-capitalization, placeholder text, and different keyboard types, such as a numeric keypad.
- **ScrollView**: This component is a generic container for scrolling multiple views and components. There can be both vertical and horizontal scrolling (by adjusting the horizontal property) for the scrollable items. If you need to render a huge or limitless list of items, you should use `FlatList`. This supports a set of special props such as *Pull to Refresh* and *Scroll loading* (lazy-loading). If your list needs to be divided into sections, then there is also a special component for this – `SectionList`.
- **Button**: React Native has advanced components that can be used to create custom buttons and other touchable components, such as `TouchableHighlight`, `TouchableOpacity`, and `TouchableWithoutFeedback`.

- **Pressable**: This gives more precise touch control with React Native version 0.63. Basically, it is a wrapper for detecting touch. It is a well-defined component that can be used instead of touchable components such as `TouchableOpacity` and `Button`.
- **Switch**: This component resembles a checkbox; however, it is presented in the form of a switch, which we are familiar with on mobile devices.

In the following chapters, we will delve deeper into common components and their properties, as well as explore new components that are rarely used. We'll also look at code examples that show how to combine components to create application interfaces.

> **Note**
>
> Detailed information about all the available components can be found at `https://reactnative.dev/docs/components-and-apis`.

Summary

In this chapter, we looked at the history of the cross-platform framework React Native and what problems it solved for companies. With it, companies can use a single universal developer team to build one business logic and apply it on all platforms simultaneously, thus saving a lot of time and money. Considering, in detail, how React Native works under the hood allows us to identify potential issues at the planning stage and resolve them.

Additionally, we started to examine React Native's basic components, and with each new chapter, we will learn more about them.

In the next chapter, you'll learn how to get started with new React Native projects.

16
Kick-Starting React Native Projects

In this chapter, you'll get up and running with React Native. Thankfully, much of the boilerplate code involved with the creation of a new project is handled for you by the command-line tools. We will look at the different CLI tools for React Native apps and create our first simple app that you will be able to upload and start right on your device.

In this chapter, we'll cover the following topics:

- Exploring React Native CLI tools
- Installing and using the Expo command-line tool
- Viewing your app on your phone
- Viewing your app on Expo Snack

Technical requirements

You can find the code files of this chapter on GitHub at `https://github.com/PacktPublishing/React-and-React-Native-4th-Edition/tree/main/Chapter16/my-project`.

Exploring React Native CLI tools

To simplify and speed up the development process, we use special command-line tools that install blank projects with application templates, dependencies, and other tools for starting, building, and testing. There are two major CLI approaches we can apply:

- React Native CLI
- Expo CLI

React Native CLI is a tool created by Meta. The project is based on the original CLI tool and has three parts: native iOS and Android projects and a React Native JavaScript app. To get started, you will need either **Xcode** or **Android Studio**. One of the main advantages of React Native CLI is its flexibility. You can connect any library with a native module or directly write code to the native parts. However, all of this requires at least a basic understanding of mobile development.

Expo CLI is just one part of the big ecosystem for developing React Native apps. Expo is a framework and a platform for universal React applications. Built around React Native and native platforms, it allows you to build, deploy, test, and rapidly iterate on iOS, Android, and web apps from a single JavaScript/TypeScript code base.

The Expo framework provides the following:

- **Expo CLI**: A command-line tool that can create blank projects, then run, build, and update them.
- **Expo Go**: An Android and iOS app for running your projects directly on your device (without having to compile and sign native apps) and sharing them across your entire team.
- **Expo Snack**: The online playground that allows you to develop React Native apps in the browser.
- **Expo Application Services (EAS)**: A set of deeply integrated cloud services for Expo and React Native applications. Apps can be compiled, signed, and uploaded to the stores using EAS in the cloud.

Expo comes with a huge number of ready-to-use features, but it also imposes limitations on projects. We cannot manually connect custom libraries with native modules since we do not have direct access to the native part of our app. Expo provides us with the most popular native libraries that we can use with Expo CLI. Expo also includes a particular version of React Native, and to update it we must wait until Expo CLI is compatible with the new React Native version.

Since Expo is useful for new developers without mobile development skills, we will use it to set up our first React Native project.

Installing and using the Expo command-line tool

The Expo command-line tool handles the creation of all of the scaffolding that your project needs to run a basic React Native application. Additionally, Expo has a couple of other tools that make running our app during development nice and straightforward. But first, we need to install the Expo command-line tool:

1. In your command-line terminal, type in the following command:

```
npm install -g expo-cli
```

2. Once this installation is complete, you'll have a new `expo` command available on your system. To start a new project, we can run the `expo init` command, as follows:

```
expo init my-project
```

3. In this case, the name of the project that will be created is `my-project`. Next, the process will ask you about your project. You should see something like this in your terminal:

```
? Choose a template: ' - Use arrow-keys. Return to
submit.
    ----- Managed workflow -----
›   blank - a minimal app as clean as an empty canvas
    blank (TypeScript) - same as blank but with
    TypeScript configuration
    tabs (TypeScript) - several example screens and
    tabs using react-navigation and TypeScript
    ----- Bare workflow -----
    Minimal - bare and minimal, just the essentials to
    get you started
```

We'll choose the blank `Managed workflow` (the default). `Managed` means that, later on, we can use Expo tools and services during development that will enable us to focus more on the application than on the complexities of developing for different mobile devices.

4. After installing all the dependencies, Expo will finish creating your project for you:

```
Extracting project files... Customizing project...
Installing dependencies...
✅ Your project is ready!
```

Now that we have created a blank React Native project, you'll learn how to launch the Expo development server on your computer and view the app on one of your devices.

Viewing your app on your phone

In order to view your React Native project on your device during development, we need to start the Expo development server:

1. In the command-line terminal, make sure that you're in the project directory:

```
cd path/to/my-project
```

2. Once you're in `my-project`, you can run the following command to start the development server:

```
npm start
```

3. This will show you some information about the developer server in the terminal:

```
Starting project at /Users/sakhniuk/React-and-React-
Native-4th-Edition-/Chapter16/my-project
Developer tools running on http://localhost:19002
Starting Metro Bundler
' Metro waiting on exp://192.168.1.233:19000
' Scan the QR code above with Expo Go (Android) or the
  Camera app (iOS)
```

4. It will also open a browser tab with the Expo developer tools for managing where the application is run, viewing logs, and other miscellaneous activities. *Figure 16.1* shows what the Expo developer tools look like:

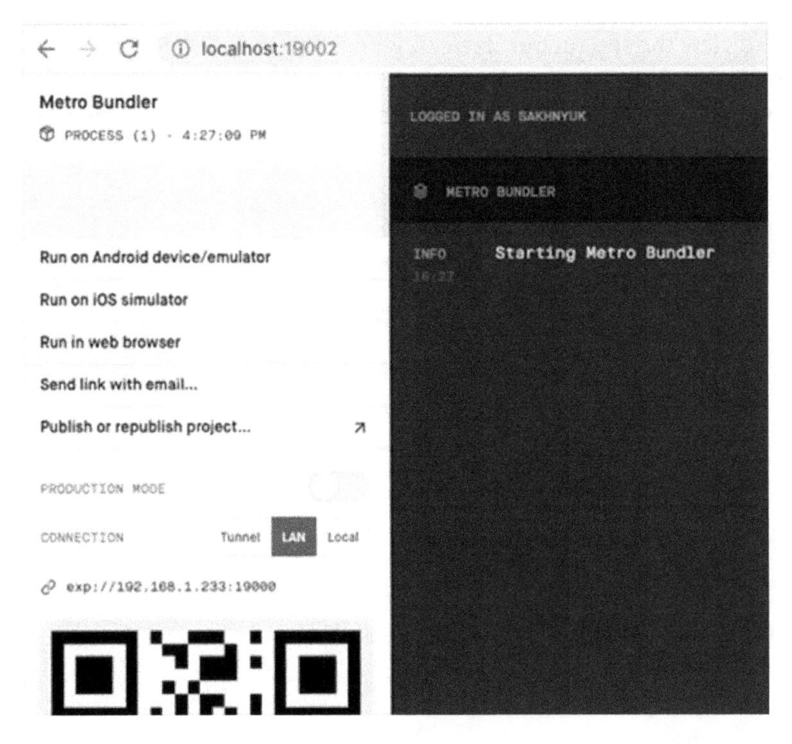

Figure 16.1 – Expo developer tools

On the right side of the screen is where you'll find logs that come from the bundler, the process that bundles your React Native code and sends it to an emulator or a physical device. In the bottom-left corner of the page is a QR code that you can scan with the camera on your device. This is how we deliver bundled React Native code to physical devices. If your device doesn't have a camera, you can click on the **Send link with email...** button.

5. In order to view the app on our devices, we need to install the **Expo Go** app. You can find it in the Play Store on Android devices or in the App Store on iOS devices. Once you have Expo installed, you can click on the **Scan QR Code** button or use the native camera on iOS devices:

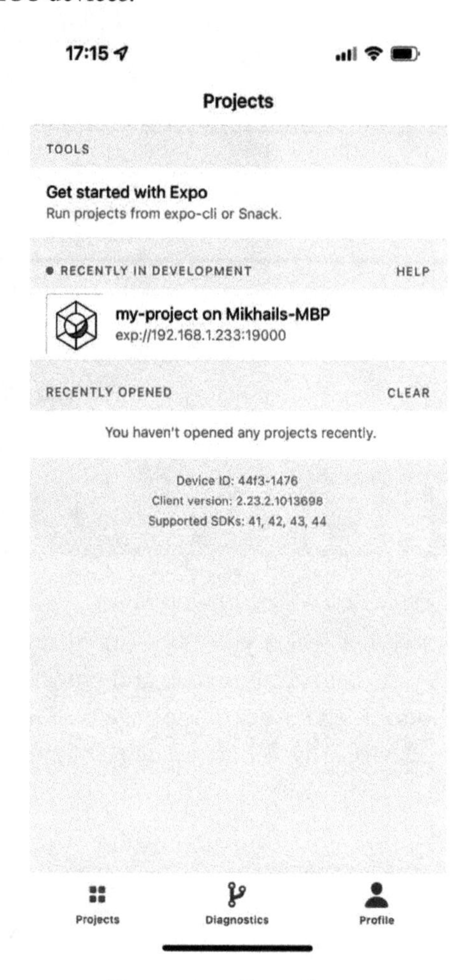

Figure 16.2 – Expo Go app

If you logged in to Expo Go and Expo CLI, you will be able to run the app without the QR code. In *Figure 16.2*, you can see the opened development session for `my-project`; if you click on it, the app will be run.

6. Once the QR code is scanned or your opened session on Expo Go is clicked, you'll notice new logs and a new connected device in the Expo UI:

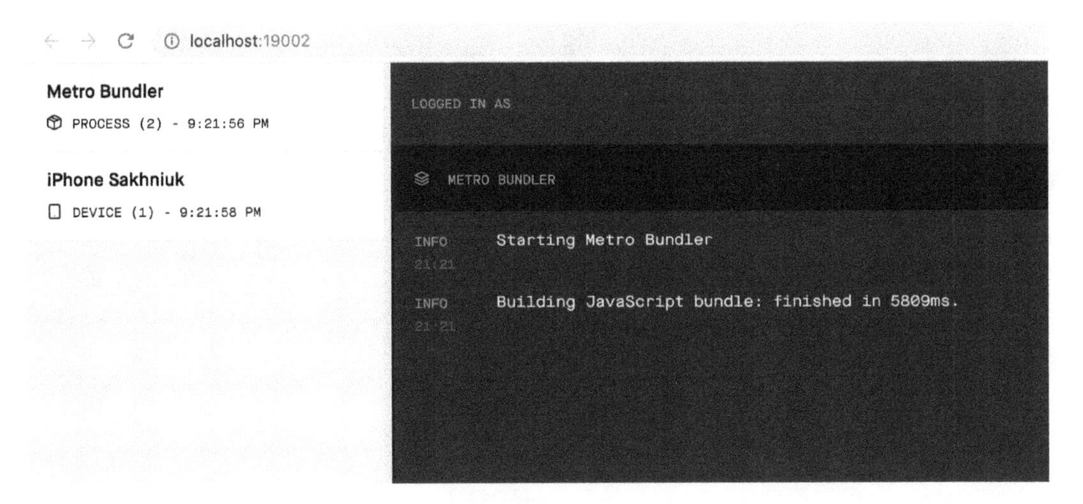

Figure 16.3 – Connected device and JavaScript build bundled

7. If you see your device listed on the left side of this screen and the logs on the right side, that indicates a JavaScript bundle has been built. You can return to your device and you should see your app running:

17:29

Open up App.js to start working on your app!

Figure 16.4 – Opened app in Expo Go

At this point, you're ready to start developing your app. In fact, you can repeat this same process if you have several physical devices that you want to work with at the same time. The best part of this Expo setup is that we get live reloading for free on our physical devices as we make code updates on our computer. Let's try this now to make sure that everything works as expected:

1. Let's open up the `App.js` file inside the `my-project` folder:

```
import { StatusBar } from 'expo-status-bar';
import { StyleSheet, Text, View } from 'react-native';

export default function App() {
  return (
    <View style={styles.container}>
      <Text>Open up App.js to start working on your
        app!</Text>
      <StatusBar style="auto" />
    </View>
  );
}

const styles = StyleSheet.create({
  container: {
    flex: 1,
    backgroundColor: '#fff',
    alignItems: 'center',
    justifyContent: 'center',
  },
});
```

2. Now let's make a small style change to make the font bold:

```
import { StatusBar } from "expo-status-bar";
import { StyleSheet, Text, View } from "react-native";

export default function App() {
  return (
    <View style={styles.container}>
      <Text style={styles.text}>
        Open up App.js to start working on your app!
      </Text>
      <StatusBar style="auto" />
    </View>
  );
}

const styles = StyleSheet.create({
  container: {
    flex: 1,
    backgroundColor: "#fff",
    alignItems: "center",
    justifyContent: "center",
  },
  text: {
    fontWeight: "bold",
  },
});
```

3. We've added a new style called `text` and applied it to the `Text` component. If you save the file and return to your device, you'll immediately see the change applied:

21:28
Open up App.js to start working on your app!

Figure 16.5 – App with updates to style of text

Now that you're able to run your apps locally on your physical devices, it's time to look at running your React Native apps on a variety of virtual device emulators using the Expo Snack service.

Viewing your app on Expo Snack

The Snack service provided by Expo is a playground for your React Native code. It lets you organize your React Native project files just like you would locally on your computer. If you end up putting something together that is worth building on, you can export your Snack. You can also create an Expo account and save your Snacks to keep working on them or to share them with others. You can find Expo Snack by the link: `https://snack.expo.dev/`.

We can create a React Native app in Expo Snack from scratch, and it will be stored in an Expo account, or we can import existing projects from a Git repository. The nice thing about importing a repository is that when you push changes to Git, your Snack will also be updated. The Git URL for the example that we've worked on in this chapter looks like this: `https://github.com/PacktPublishing/React-and-React-Native-4th-Edition-/tree/main/Chapter16/my-project`.

We can click on the **Import git repository** button in the Snack project menu and paste in this URL:

Figure 16.6 – Importing a Git repository to Expo Snack

Once the repository is imported and the Snack is saved, you'll get an updated Snack URL that reflects the Git repository location. For example, the Snack URL from this chapter looks like this: `https://snack.expo.dev/@sakhnyuk/2a2429`.

If you open this URL, the Snack interface will load and you can make changes to the code to test things out before running them. The main advantage of Snack is the ability to easily run them on virtualized devices. The controls to run your app on a virtual device can be found on the right side of the UI and look like this:

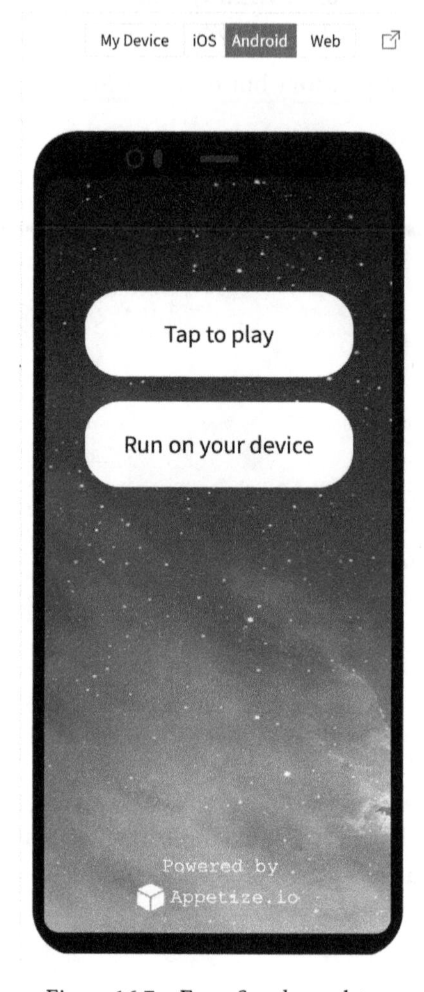

Figure 16.7 – Expo Snack emulator

The top control above the image of the phone controls which device type to emulate: **Android**, **iOS**, or **Web**. The **Tap to play** button will launch the selected virtual device. The **Run on your device** button allows you to run the app in Expo Go using the QR code approach.

Here's what our app looks like on a virtual iOS device:

9:14
Open up App.js to start working on your app!

Figure 16.8 – Expo Snack iOS emulator

And here's what our app looks like on a virtual Android device:

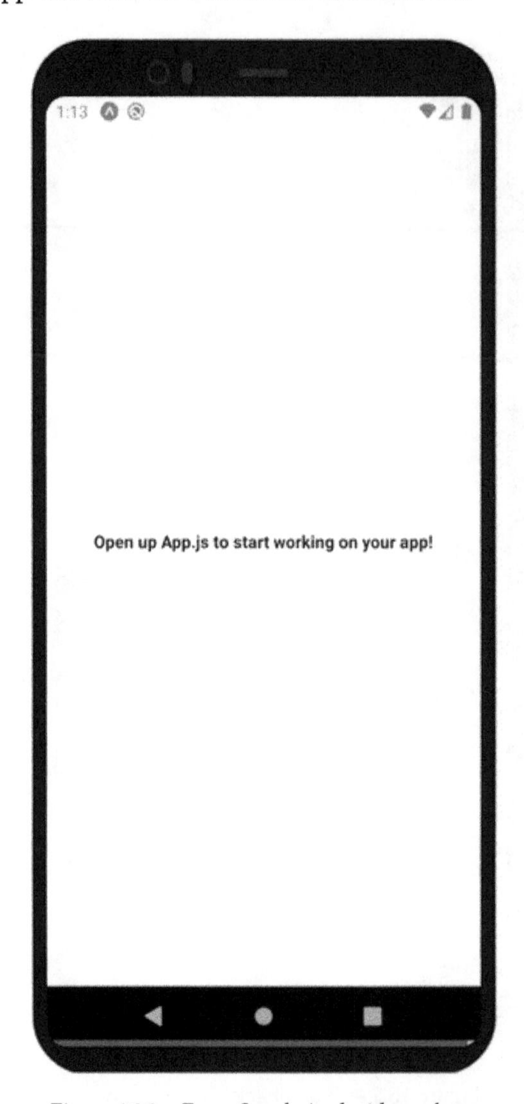
1:13
Open up App.js to start working on your app!

Figure 16.9 – Expo Snack Android emulator

This app only displays text and applies some styles to it, so it looks pretty much identical on different platforms. As we make our way through the React Native chapters in this book, you'll see how useful a tool such as Snack is for making comparisons between the two platforms to understand the difference between them.

Summary

In this chapter, you learned how to kick-start a React Native project using the Expo command-line tool. First, you learned how to install the Expo tool. Then, you learned how to initialize a new React Native project. Next, you started the Expo development server and learned about the various parts of the development server UI.

In particular, you learned how to connect the development server with the Expo app on any device that you want to test your app on. Expo also has the Snack service, which lets us experiment with snippets of code or entire Git repositories. You learned how to import a repository and run it on virtual iOS and Android devices.

In the next chapter, we'll look at how to build responsive layouts in our React Native apps.

17
Building Responsive Layouts with Flexbox

In this chapter, you'll get a feel for what it's like to lay components out on the screen of mobile devices. Thankfully, React Native polyfills many CSS properties that you might have used in the past to implement page layouts in web applications.

Before you dive into implementing layouts, you'll get a brief introduction to Flexbox and using CSS style properties in React Native apps – it's not quite what you're used to with regular CSS style sheets. Then, you'll implement several React Native layouts using Flexbox.

Here's the list of topics that we'll cover in this chapter:

- Introducing Flexbox
- Introducing React Native styles
- Using the Styled Components library
- Building Flexbox layouts

Technical requirements

You can find the code files present in this chapter on GitHub at `https://github.com/PacktPublishing/React-and-React-Native-4th-Edition/tree/main/Chapter17`.

Introducing Flexbox

Before the flexible box layout model was introduced to CSS, the various approaches used to build layouts felt hacky and were prone to errors. Flexbox fixes this by abstracting many of the properties that you would normally have to provide in order to make the layout work.

In essence, the Flexbox model is exactly what it sounds like: a box model that's flexible. That's the beauty of Flexbox – its simplicity. You have a box that acts as a container, and you have child elements within that box. Both the container and the child elements are flexible in how they're rendered on the screen, as illustrated here:

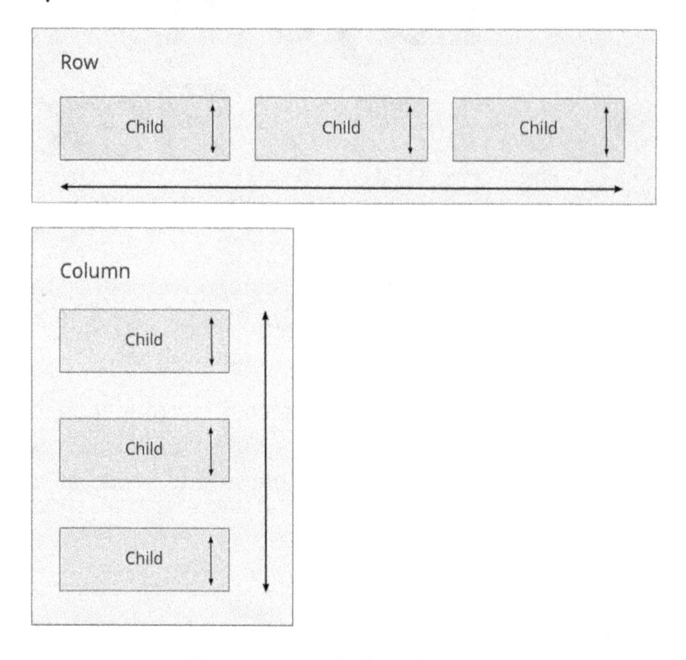

Figure 17.1 – Flexbox elements

Flexbox containers have a direction, either **Column** (up/down) or **Row** (left/right). This actually confused me when I was first learning Flexbox; my brain refused to believe that rows move from left to right. Rows stack on top of one another! The key thing to remember is that it's the direction that the box flexes, not the direction that boxes are placed on the screen.

> **Important Note**
>
> For a more in-depth treatment of Flexbox concepts, refer to `https://css-tricks.com/snippets/css/a-guide-to-Flexbox`.

Now that we've covered the basics of Flexbox layouts at a high level, it's time to learn how styles in React Native applications work.

Introducing React Native styles

It's time to implement your first React Native app, beyond the boilerplate that's generated by `create-react-native-app`. I want to make sure that you feel comfortable using React Native style sheets before you start implementing Flexbox layouts in the next section.

Here's what a React Native style sheet looks like:

```
import { Platform, StyleSheet, StatusBar } from "react-native";
export default StyleSheet.create({
  container: {
    flex: 1,
    justifyContent: "center",
    alignItems: "center",
    backgroundColor: "ghostwhite",
    ...Platform.select({
      ios: { paddingTop: 20 },
      android: { paddingTop: StatusBar.currentHeight },
    }),
  },
  box: {
    width: 100,
    height: 100,
    justifyContent: "center",
    alignItems: "center",
    backgroundColor: "lightgray",
  },
  boxText: {
    color: "darkslategray",
    fontWeight: "bold",
```

```
  },
});
```

This is a JavaScript module, not a CSS module. If you want to declare React Native styles, you need to use plain objects. Then, you call `StyleSheet.create()` and export this from the style module. Note that style names are pretty similar to the web CSS, except that they are written in camel case; for example, `justifyContent` rather than `justify-content`.

As you can see, this style sheet has three styles: `container`, `box`, and `boxText`. Within the container style, there's a call to `Platform.select()`:

```
...Platform.select({
ios: { paddingTop: 20 },
android: { paddingTop: StatusBar.currentHeight }
})
```

This function will return different styles based on the platform of the mobile device. Here, you're handling the top padding of the top-level container view. You'll probably use this code in most of your apps to make sure that your React components don't render underneath the status bar of the device. Depending on the platform, the padding will require different values. If it's iOS, `paddingTop` is `20`. If it's Android, `paddingTop` will be the value of `StatusBar.currentHeight`.

> **Important Note**
>
> The preceding `Platform.select()` code is an example of a case where you need to implement a workaround for differences in the platform. For example, if `StatusBar.currentHeight` was available on iOS and Android, you wouldn't need to call `Platform.select()`.

Let's see how these styles are imported and applied to React Native components:

```
import React from "react";
import { Text, View } from "react-native";
import styles from "./styles";
export default function App() {
  return (
    <View style={styles.container}>
      <View style={styles.box}>
        <Text style={styles.boxText}>I'm in a box</Text>
```

```
      </View>
    </View>
  );
}
```

The styles are assigned to each component via the `style` property. You're trying to render a box with some text in the middle of the screen. Let's make sure that this looks as we expect it to.

Figure 17.2 – Box in a middle of a screen

We have found out how to apply styles to components using a built-in module, but there is more than one way to define styles. We also have the option to write CSS in React Native. Let's quickly go through it.

Using the Styled Components library

Styled components is a CSS-in-JS library that styles React Native components using plain CSS. With this approach, you don't need to define style classes via objects and provide style props. The CSS itself is determined via tagged template literals provided by `styled-components`.

To install `styled-components`, run this command in your project:

```
npm install --save styled-components
```

Let's try to rewrite components from the *Introducing React Native styles* section. Here is what our `Box` component will look like:

```
import styled from "styled-components/native";

const Box = styled.View'
  width: 100px;
  height: 100px;
  justify-content: center;
  align-items: center;
  background-color: lightgray;
';

const BoxText = styled.Text'
  color: darkslategray;
  font-weight: bold;
';
```

In this example, we've got two components, `Box` and `BoxText`. Now we can use them as usual, but without any other additional styling props:

```
const App = () => {
  return (
    <Box>
      <BoxText>I'm in a box</BoxText>
    </Box>
  );
};
```

In further sections, I will use `StyleSheet` objects, but if you want to learn more about `styled-components`, you can read more here: `https://styled-components.com/`.

Perfect! Now that you have an idea of how to set styles on React Native elements, it's time to start creating some screen layouts.

Building Flexbox layouts

In this section, you'll learn about several potential layouts that you can use in your React Native applications. I want to stay away from the idea that one layout is better than another. Instead, I'll show you how powerful the Flexbox layout model is for mobile screens so that you can design the kind of layout that best suits your application.

Simple three-column layout

To start things off, let's implement a simple layout with three sections that flex in the direction of the column (top to bottom). We'll look at the result we are aiming for first.

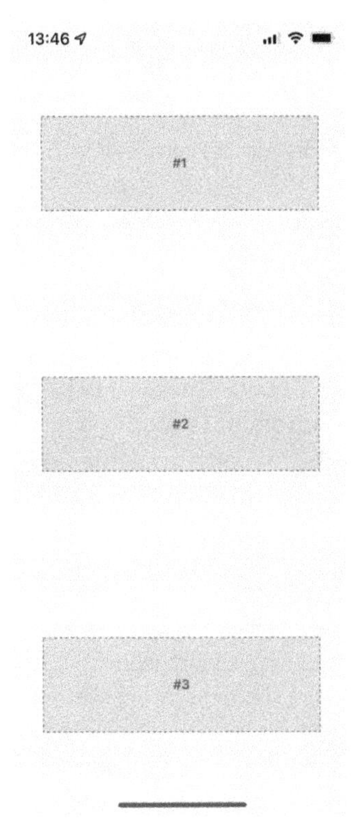

Figure 17.3 – Simple three-column layout

The idea, in this example, is that you've styled and labeled the three screen sections so that they stand out. In other words, these components wouldn't necessarily have any styling in a real application since they're used to arrange other components on the screen.

Now, let's take a look at the components used to create this screen layout:

```
import React from "react";
import { Text, View } from "react-native";
import styles from "./styles";

export default function App() {
  return (
    <View style={styles.container}>
      <View style={styles.box}>
        <Text style={styles.boxText}>#1</Text>
      </View>
      <View style={styles.box}>
        <Text style={styles.boxText}>#2</Text>
      </View>
      <View style={styles.box}>
        <Text style={styles.boxText}>#3</Text>
      </View>
    </View>
  );
}
```

The container view (the outermost `<View>` component) is the column and the child views are the rows. The `<Text>` component is used to label each row. In terms of HTML elements, `<View>` is similar to a `<div>` element, while `<Text>` is similar to a `<p>` element.

> **Important Note**
>
> Maybe this example could have been called a *three-row layout* since it has three rows. But, at the same time, the three layout sections are flexing in the direction of the column that they're in. Use the naming convention that makes the most conceptual sense to you.

Now, let's take a look at the styles used to create this layout:

```
import { Platform, StyleSheet, StatusBar } from "react-native";

export default StyleSheet.create({
  container: {
    flex: 1,
    flexDirection: "column",
    alignItems: "center",
    justifyContent: "space-around",
    backgroundColor: "ghostwhite",
    ...Platform.select({
      ios: { paddingTop: 20 },
      android: { paddingTop: StatusBar.currentHeight }
    })
  },

  box: {
    width: 300,
    height: 100,
    justifyContent: "center",
    alignItems: "center",
    backgroundColor: "lightgray",
    borderWidth: 1,
    borderStyle: "dashed",
    borderColor: "darkslategray"
  },

  boxText: {
    color: "darkslategray",
    fontWeight: "bold"
  }
});
```

The `flex` and `flexDirection` properties of `container` enable the layout of the rows to flow from top to bottom. The `alignItems` and `justifyContent` properties align the child elements to the center of the container and add space around them, respectively.

Let's see how this layout looks when you rotate the device from a portrait orientation to a landscape orientation:

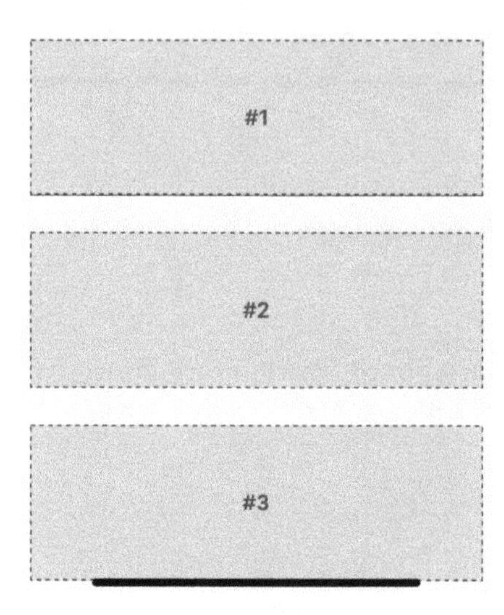

Figure 17.4 – Landscape orientation

Flexbox automatically figured out how to preserve the layout for you. However, you can improve on this a little bit. For example, the landscape orientation now has a lot of wasted space to the left and right. You could create your own abstraction for the boxes that you're rendering. In the following section, we'll improve on this layout.

Improved three-column layout

There are a few things that I think you can improve on from the last example. Let's fix the styles so that the children of the Flexbox could stretch to take advantage of the available space. Do you remember, in the last example, when you rotated the device from a portrait orientation to a landscape orientation? There was a lot of wasted space. It would be nice to have the components automatically adjust themselves. Here's what the new style module looks like:

```
import { Platform, StyleSheet, StatusBar } from "react-native";

export default StyleSheet.create({
  container: {
```

```
    flex: 1,
    flexDirection: "column",
    backgroundColor: "ghostwhite",
    justifyContent: "space-around",
    ...Platform.select({
      ios: { paddingTop: 20 },
      android: { paddingTop: StatusBar.currentHeight },
    }),
  },

  box: {
    height: 100,
    justifyContent: "center",
    alignSelf: "stretch",
    alignItems: "center",
    backgroundColor: "lightgray",
    borderWidth: 1,
    borderStyle: "dashed",
    borderColor: "darkslategray",
  },

  boxText: {
    color: "darkslategray",
    fontWeight: "bold",
  },
});
```

The key change here is the `alignSelf` property. This tells elements with the `box` style to change their width or height (depending on the `flexDirection` of their container) to fill space. Also, the `box` style no longer defines a `width` property because this will be computed on the fly now. Here's what the sections look like in portrait mode:

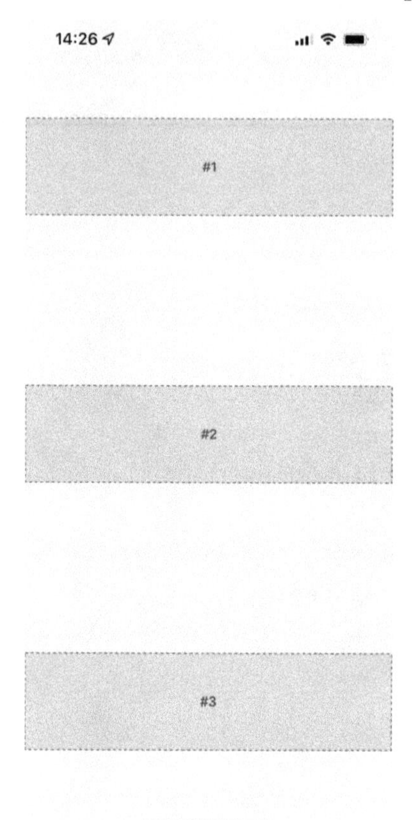

Figure 17.5 – Improved three-column layout in portrait orientation

Now, each section takes the full width of the screen, which is exactly what you want to happen. The issue of wasted space was actually more prevalent in landscape orientation, so let's rotate the device and see what happens to these sections now.

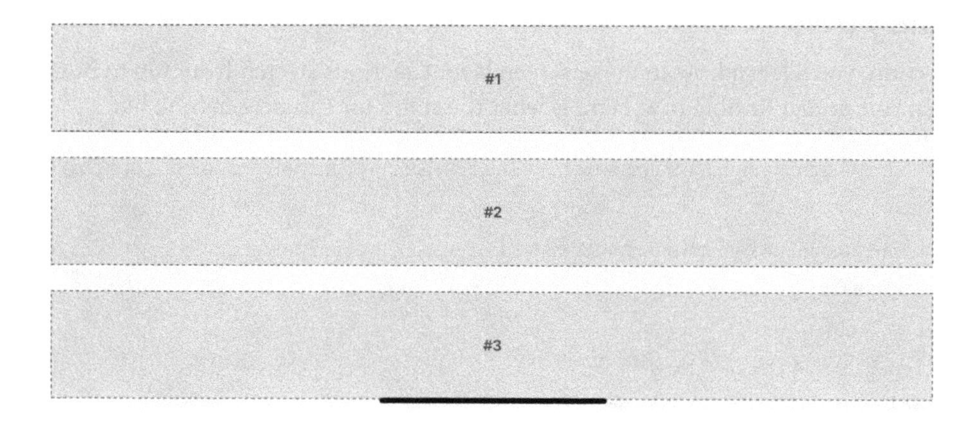

Figure 17.6 – Improved three-column layout in landscape orientation

Now your layout is utilizing the entire width of the screen, regardless of orientation. Lastly, let's implement a proper `Box` component that can be used by `App.js` instead of having repetitive style properties in place. Here's what the `Box` component looks like:

```
import React from "react";
import { PropTypes } from "prop-types";
import { View, Text } from "react-native";
import styles from "./styles";

export default function Box({ children }) {
  return (
    <View style={styles.box}>
      <Text style={styles.boxText}>{children}</Text>
    </View>
  );
}

Box.propTypes = {
  children: PropTypes.node.isRequired,
};
```

You now have the beginnings of a nice layout. Next, you'll learn about flexing in the other direction – left to right.

Flexible rows

In this section, you'll learn how to make screen layout sections stretch from top to bottom. To do this, you need a flexible row. Here is what the styles for this screen look like:

```
import { Platform, StyleSheet, StatusBar } from "react-native";

export default StyleSheet.create({
  container: {
    flex: 1,
    flexDirection: "row",
    backgroundColor: "ghostwhite",
    alignItems: "center",
    justifyContent: "space-around",
    ...Platform.select({
      ios: { paddingTop: 20 },
      android: { paddingTop: StatusBar.currentHeight },
    }),
  },

  box: {
    width: 100,
    justifyContent: "center",
    alignSelf: "stretch",
    alignItems: "center",
    backgroundColor: "lightgray",
    borderWidth: 1,
    borderStyle: "dashed",
    borderColor: "darkslategray",
  },

  boxText: {
    color: "darkslategray",
    fontWeight: "bold",
  },
});
```

Here's the App component, using the same Box component that you implemented in the previous section:

```
import React from "react";
import { Text, View, StatusBar } from "react-native";
import styles from "./styles";
import Box from "./Box";

export default function App() {
  return (
    <View style={styles.container}>
      <Box>#1</Box>
      <Box>#2</Box>
    </View>
  );
}
```

Here's what the resulting screen looks like in portrait mode:

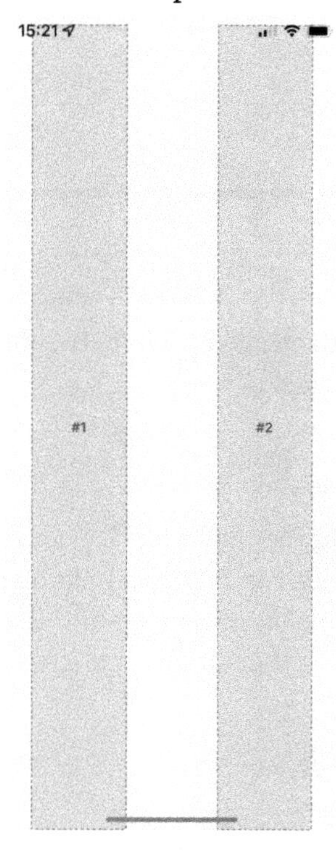

Figure 17.7 – Flexible rows in portrait orientation

The two columns stretch all the way from the top of the screen to the bottom because of the `alignSelf` property, which doesn't actually specify which direction to stretch in. The two `Box` components stretch from top to bottom because they're displayed in a flex row. Note how the spacing between these two sections goes from left to right? This is because of the container's `flexDirection` property, which has a value of `row`.

Now, let's see how this flex direction impacts the layout when the screen is rotated to a landscape orientation.

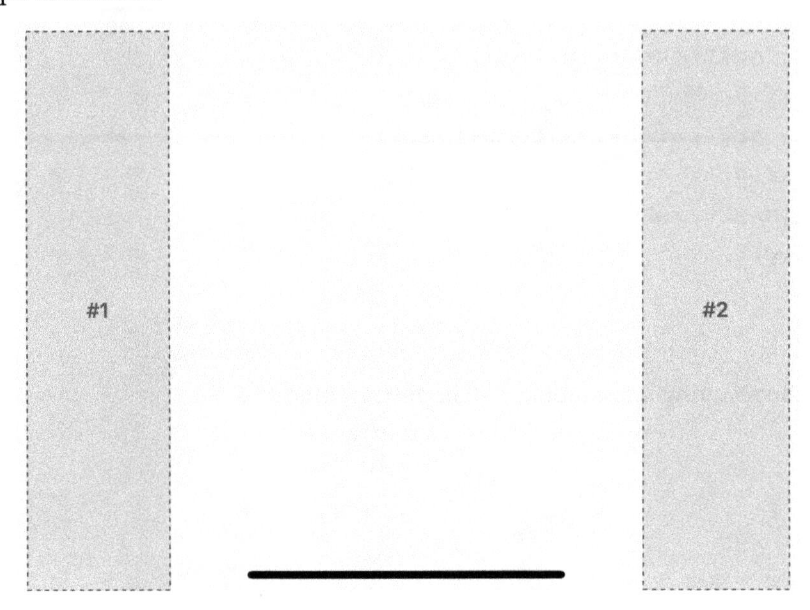

Figure 17.8 – Flexible rows in landscape orientation

Since Flexbox has a `justifyContent` style property value of `space-around`, space is added proportionally to the left, the right, and in between the sections. In the following section, you'll learn about flexible grids.

Flexible grids

Sometimes, you need a screen layout that flows like a grid. For example, what if you have several sections that are the same width and height, but you're not sure how many of these sections will be rendered? Flexbox makes it easy to build a row that flows from left to right until the end of the screen is reached. Then, it automatically continues rendering elements from left to right on the next row.

Here's an example layout in portrait mode:

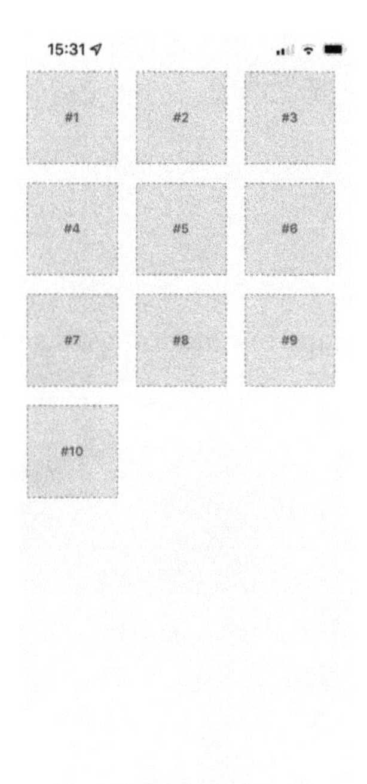

Figure 17.9 – Flexible grids in portrait orientation

The beauty of this approach is that you don't need to know in advance how many columns are in a given row. The dimensions of each child determine what will fit in a given row.

To see the styles used to create this layout, you can follow this link: `https://github.com/PacktPublishing/React-and-React-Native-4th-Edition/blob/main/Chapter17/flexible-grids/styles.js`.

Here's the `App` component that renders each section:

```
import React from "react";
import { View, StatusBar } from "react-native";
import styles from "./styles";
import Box from "./Box";

const boxes = new Array(10).fill(null).map((v, i) => i + 1);

export default function App() {
  return (
    <View style={styles.container}>
```

```
      <StatusBar hidden={false} />
      {boxes.map((i) => (
        <Box key={i}>#{i}</Box>
      ))}
    </View>
  );
}
```

Lastly, let's make sure that the landscape orientation works with this layout:

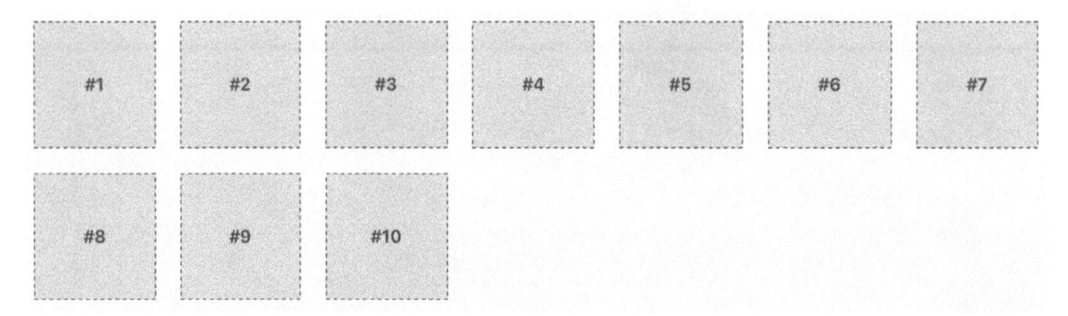

Figure 17.10 – Flexible grids in landscape orientation

> **Important Note**
> You may have noticed that there's some superfluous space on the right side. Remember, these sections are only visible in this book because we want them to be visible. In a real app, they're just grouping other React Native components. However, if the space to the right of the screen becomes an issue, play around with the margin and the width of the child components.

Now that you have an understanding of how flexible grids work, we'll look at flexible rows and columns next.

Flexible rows and columns

In this final section of the chapter, you'll learn how to combine rows and columns to create a sophisticated layout for your app. For example, sometimes, you need the ability to nest columns within rows or rows within columns. To see the `App` component of an application that nests columns within rows, you can follow this link: `https://github.com/PacktPublishing/React-and-React-Native-4th-Edition/blob/main/Chapter17/flexible-rows-and-columns/App.js`.

You've created abstractions for the layout pieces (`<Row>` and `<Column>`) and the content piece (`<Box>`). Let's see what this screen looks like:

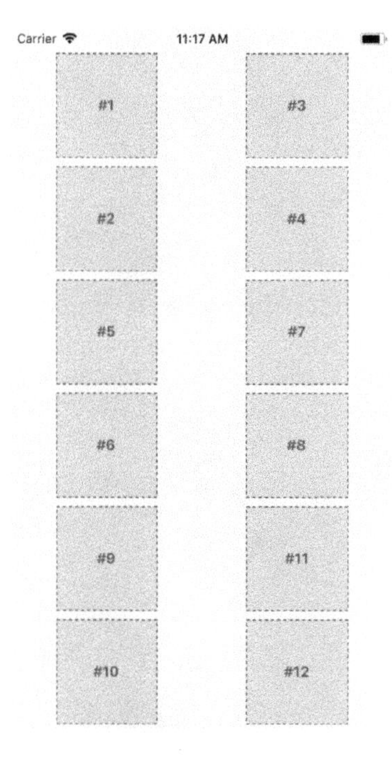

Figure 17.11 – Flexible rows and columns

This layout probably looks familiar because you've done it in the *Flexible grids* section. The key difference as compared to *Figure 17.9* is in how these content sections are ordered. For example, `#2` doesn't go to the right of `#1`, it goes below it. This is because we've placed `#1` and `#2` in `<Column>`. The same happens with `#3` and `#4`. These two columns are placed in a row. Then, the next row begins, and so on.

This is just one of many possible layouts that you can achieve by nesting row Flexboxes and column Flexboxes. Let's take a look at the `Row` component now:

```
import React from "react";
import PropTypes from "prop-types";
import { View, Text } from "react-native";
import styles from "./styles";

export default function Box({ children }) {
  return (
    <View style={styles.box}>
```

```
      <Text style={styles.boxText}>{children}</Text>
    </View>
  );
}

Box.propTypes = {
  children: PropTypes.node.isRequired,
};
```

This component applies the `row` style to the `<View>` component. The end result is cleaner JSX markup in the `App` component when creating a complex layout. Finally, let's look at the `Column` component:

```
import React from "react";
import PropTypes from "prop-types";
import { View } from "react-native";
import styles from "./styles";

export default function Column({ children }) {
  return <View style={styles.column}>{children}</View>;
}

Column.propTypes = {
  children: PropTypes.node.isRequired,
};
```

This looks just like the `Row` component, just with a different style applied to it. It also serves the same purpose as `Row` – to enable simpler JSX markup for layouts in other components.

Summary

This chapter introduced you to styles in React Native. Though you can use many of the same CSS style properties that you're used to, the CSS style sheets used in web applications look very different. Namely, they're composed of plain JavaScript objects.

Then, you learned how to work with the main React Native layout mechanism – Flexbox. This is the preferred way of laying out most web applications these days, so it makes sense to be able to reuse this approach in a Native app. You created several different layouts, and you saw how they looked in portrait and landscape orientation.

In the next chapter, you'll start implementing navigation for your app.

Further reading

Refer to the following links for more information:

- Layout with Flexbox: `https://reactnative.dev/docs/flexbox`
- `StatusBar`: `https://reactnative.dev/docs/statusbar`
- `StyleSheet`: `https://reactnative.dev/docs/stylesheet`

18
Navigating Between Screens

The focus of this chapter is on navigating between the screens that make up your React Native application. Navigation in native apps is slightly different than navigation on web apps – mainly because there isn't any notion of a URL that the user is aware of. In prior versions of React Native, there were primitive navigator components that you could use to control the navigation between screens. There were a number of challenges with these components that resulted in more code to accomplish basic navigation tasks.

More recent versions of React Native encourage you to use the `react-navigation` package, which will be the focus of this chapter, even though there are several other options. You'll learn about navigation basics, passing parameters to screens, changing header content, using tab and drawer navigation, and handling state with navigation.

We'll cover the following topics in this chapter:

- Navigation basics
- Route parameters
- The navigation header
- Tab and drawer navigation

Technical requirements

You can find the code files for this chapter on GitHub at `https://github.com/PacktPublishing/React-and-React-Native-4th-Edition/tree/main/Chapter18`.

Navigation basics

Let's start off with the basics of moving from one page to another using the `react-navigation` package.

Before starting, you should install the `react-navigation` package to our project and some additional dependencies related to the example:

```
npm install @react-navigation/native
```

Then, install native dependencies using `expo`:

```
expo install react-native-screens react-native-safe-area-
context
```

The preceding installation steps will be required for each example in this chapter, but we need to add one more package related to the stack navigator:

```
npm install @react-navigation/native-stack
```

Now, we are ready to develop navigation. Here's what the `App` component looks like:

```
import * as React from "react";
import { NavigationContainer } from
  "@react-navigation/native";
import { createNativeStackNavigator } from
  "@react-navigation/native-stack";
import Home from "./Home";
import Settings from "./Settings";

const Stack = createNativeStackNavigator();

export default function App() {
  return (
    <NavigationContainer>
      <Stack.Navigator>
```

```
        <Stack.Screen name="Home" component={Home} />
        <Stack.Screen name="Settings" component={Settings}
          />
      </Stack.Navigator>
    </NavigationContainer>
  );
}
```

`createNativeStackNavigator()` is a function that sets up your navigation. It returns an object with two properties, the `Screen` and `Navigator` components, that are used for configuring the stack navigator.

The first argument to this function maps to the screen components that can be navigated. The second argument is for more general navigation options – in this case, you're telling the navigator that `Home` should be the default screen component that's rendered. The `<NavigationContainer>` component is necessary so that the screen components get all of the navigation properties that they need.

Here's what the `Home` component looks like:

```
import React from "react";
import { View, Text, Button, StatusBar } from
  "react-native";
import styles from "./styles";

export default function Home({ navigation }) {
  return (
    <View style={styles.container}>
      <StatusBar barStyle="dark-content" />
      <Text>Home Screen</Text>
      <Button
        title="Settings"
        onPress={() => navigation.navigate("Settings")}
      />
    </View>
  );
}
```

This is your typical functional React component. You can use a class-based component here, but there's no need, since there are no life cycle methods or state. It renders a `View` component where the container style is applied. This is followed by a `Text` component that labels the screen followed by a `Button` component. A screen can be anything you want – it's just a regular React Native component. The navigator component handles the routing and the transitions between screens for you.

The `onPress` handler for this button navigates to the `Settings` screen when clicked. This is done by calling `navigation.navigate('Settings')`. The `navigation` property is passed to your screen component by `react-navigation` and contains all of the routing functionality you need. In contrast to working with URLs in React web apps, here you call navigator API functions and pass them the names of screens.

Let's take a look at the `Settings` component:

```
import React from "react";
import { View, Text, Button, StatusBar } from
  "react-native";
import styles from "./styles";

export default function Settings({ navigation }) {
  return (
    <View style={styles.container}>
      <StatusBar barStyle="dark-content" />
      <Text>Settings Screen</Text>
      <Button title="Home" onPress={() =>
        navigation.navigate("Home")} />
    </View>
  );
}
```

This component is just like the `Home` component, except with different text, and when the button is clicked, you're taken back to the **Home** screen.

Here's what the **Home** screen looks like:

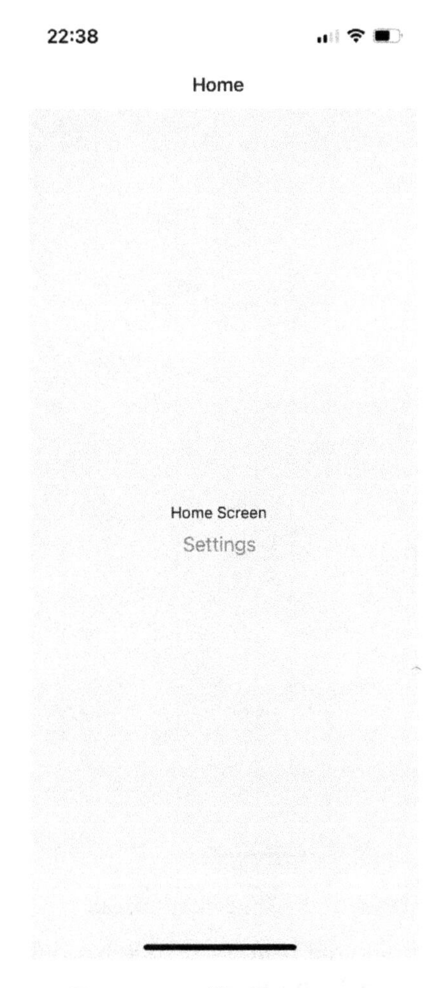

Figure 18.1 – The Home screen

If you click the **Settings** button, you'll be taken to the **Settings** screen, which looks like this:

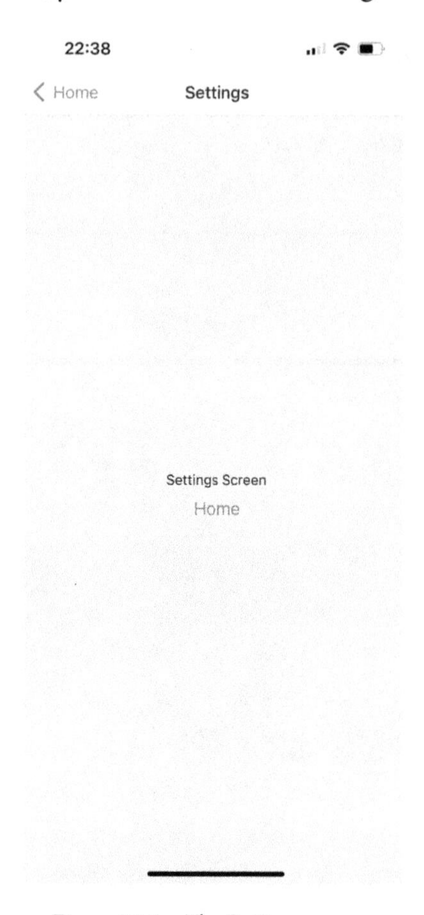

Figure 18.2 – The Settings screen

This screen looks almost identical to the **Home** screen. It has different text and a different button that will take you back to the **Home** screen when clicked. However, there's another way to get back to the **Home** screen. Take a look at the top of the screen, where you'll notice a white navigation bar. On the left side of the navigation bar, there's a back arrow. This works just like the back button in a web browser and will take you back to the previous screen. What's nice about `react-navigation` is that it takes care of rendering this navigation bar for you.

> **Important Note**
>
> With this navigation bar in place, you don't have to worry about how your layout styles impact the status bar. You only need to worry about the layout of each of your screens.

If you run this app on Android, you'll see the same back button in the navigation bar. But you can also use the standard back button found outside of the app on most Android devices.

In the next section, you'll learn how to pass parameters to your routes.

Route parameters

When you develop React web applications, some of your routes have dynamic data in them. For example, you can link to a details page, and within that URL, you'll have some sort of identifier. The component then has what it needs to render specific detailed information. The same concept exists within `react-navigation`. Instead of just specifying the name of the screen that you want to navigate to, you can pass along additional data.

Let's take a look at route parameters in action:

1. We'll start with the `App` component:

```
import { NavigationContainer } from "@react-navigation/
native";
import { createNativeStackNavigator } from "@react-
navigation/native-stack";
import Home from "./Home";
import Details from "./Details";

const Stack = createNativeStackNavigator();

export default function App() {
  return (
    <NavigationContainer>
      <Stack.Navigator>
        <Stack.Screen name="Home" component={Home} />
        <Stack.Screen name="Details"
          component={Details} />
      </Stack.Navigator>
    </NavigationContainer>
  );
}
```

This looks just like the example in the *Navigation basics* section, except instead of a `Settings` page, there's a `Details` page. This is the page that you want to pass data to dynamically so that it can render the appropriate information.

2. Next, let's take a look at the `Home` screen component:

```
import React from "react";
import { View, Text, Button, StatusBar } from
  "react-native";
import styles from "./styles";

export default function Home({ navigation }) {
  return (
    <View style={styles.container}>
      <StatusBar barStyle="dark-content" />
      <Text>Home Screen</Text>
      <Button
        title="First Item"
        onPress={() => navigation.navigate("Details",
          { title: "First Item" })}
      />
      <Button
        title="Second Item"
        onPress={() => navigation.navigate("Details",
          { title: "Second Item" })}
      />
      <Button
        title="Third Item"
        onPress={() => navigation.navigate("Details",
          { title: "Third Item" })}
      />
    </View>
  );
}
```

The `Home` screen has three `Button` components, and each navigates to the `Details` screen. Note that in the `navigation.navigate()` calls, in addition to the screen name, each has a second argument. These arguments are objects that contain specific data, which is passed to the `Details` screen.

3. Next, let's take a look at the `Details` screen and see how it consumes these route parameters:

```
import React from "react";
import { View, Text, StatusBar } from "react-native";
import styles from "./styles";

export default function ({ route }) {
  const { title } = route.params;
  return (
    <View style={styles.container}>
      <StatusBar barStyle="dark-content" />
      <Text>{title}</Text>
    </View>
  );
}
```

Although this example is only passing one `title` parameter, you can pass as many parameters to the screen as you need to. You can access these parameters using the `params` value of the `route` prop to look up the value.

4. Here's what the **Home** screen looks like when rendered:

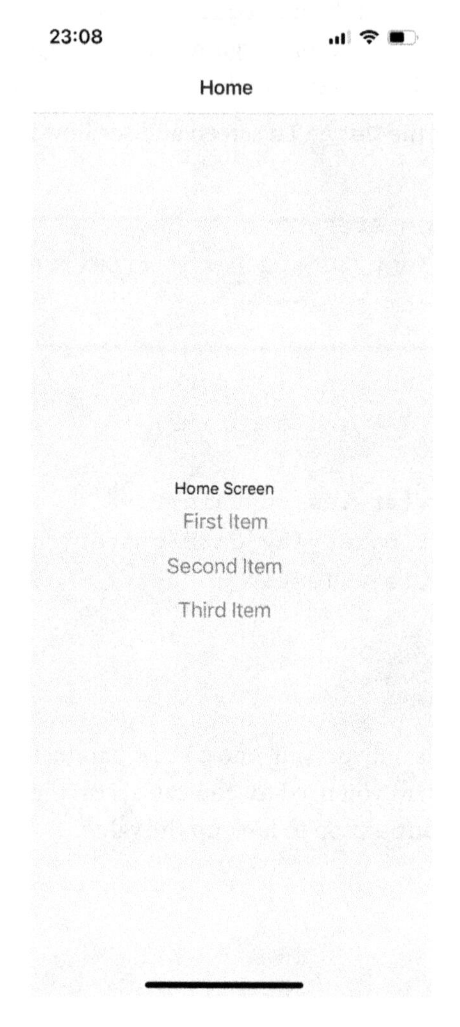

Figure 18.3 – The Home screen

5. If you click on the **First Item** button, you'll be taken to the **Details** screen that is rendered using the route parameter data:

Figure 18.4 – The Details screen

You can click the back button in the navigation bar to get back to the **Home** screen. If you click on any of the other buttons on the **Home** screen, you'll be taken back to the **Details** screen with updated data. Route parameters are necessary to avoid having to write duplicate components. You can think of passing parameters to `navigator.navigate()` as passing props to a React component.

In the following section, you'll learn how to populate navigation section headers with content.

The navigation header

The navigation bars that you've created so far in this chapter have been sort of plain. That's because you haven't configured them to do anything, so `react-navigation` will just render a plain bar with a back button. Each screen component that you create can configure specific navigation header content.

Let's build on the example discussed in the *Route parameters* section, which used buttons to navigate to a details page:

1. The `App` component has major updates, so let's take a look at it:

```
import { Button } from "react-native";
import { NavigationContainer } from
  "@react-navigation/native";
import { createNativeStackNavigator } from
  "@react-navigation/native-stack";
import Home from "./Home";
import Details from "./Details";

const Stack = createNativeStackNavigator();

export default function App() {
  return (
    <NavigationContainer>
      <Stack.Navigator>
        <Stack.Screen name="Home" component={Home} />
        <Stack.Screen
          name="Details"
          component={Details}
          options={({ route }) => ({
            headerRight: () => {
              return (
                <Button
                  title="Buy"
                  onPress={() => {}}
```

```
                    disabled={route.params.stock === 0}
                  />
                );
              },
            })}
          />
        </Stack.Navigator>
      </NavigationContainer>
    );
}
```

The Screen component accepts the options prop as an object or function to provide additional screen properties.

The headerRight option is used to add a Button component to the right side of the navigation bar. This is where the stock parameter comes into play. If this value is 0 because there isn't anything in stock, you want to disable the Buy button.

In our case, we pass options as a function and read the stock screen params to disable the button. This is one of several ways to pass options to the Screen component. We'll apply another way to the Details component.

2. To understand how the stock props have been passed, take a look at the Home component here: https://github.com/PacktPublishing/React-and-React-Native-4th-Edition-/blob/main/Chapter18/navigation-header/Home.js.

   The first thing to note is that each button is passing more route parameters to the Details component – content and stock. You'll see why in a moment.

3. Next, let's take a look at the Details component:

```
import React from "react";
import { View, Text, StatusBar } from "react-native";
import styles from "./styles";

export default function Details({ route, navigation }) {
  const { content, title } = route.params;

  React.useLayoutEffect(() => {
    navigation.setOptions({ title });
  }, []);
```

```
  return (
    <View style={styles.container}>
      <StatusBar barStyle="dark-content" />
      <Text>{content}</Text>
    </View>
  );
}
```

This time, the `Details` component renders the content route parameter. As with the `App` component, we add additional options to the screen. In this case, we update screen options using the `navigation.setOptions()` method. To customize a header, we can also add `title` to that screen via the `App` component.

4. Let's see how all of this works now, starting with the **Home** screen:

Figure 18.5 – The Home screen

There is now header text in the navigation bar, which is set by the `name` property in the `Screen` component.

5. Next, try clicking on the **First Item** button:

Figure 18.6 – The First Item screen

The title in the navigation bar is set based on the `title` parameter that's passed to the `Details` component using the `navigation.setOptions()` method. The **Buy** button that's rendered on the right side of the navigation bar is rendered by the `options` property in the `Screen` component placed in the `App` component. It's enabled because the `stock` parameter value is `1`.

6. Now, try returning to the **Home** screen and clicking on the **Second Item** button:

Figure 18.7 – The Second Item screen

The title and the page content both reflect the new parameter values passed to `Details`, but so does the **Buy** button. It is in a disabled state because the stock parameter value was `0`, meaning that it can't be bought.

Now that you've learned how to use navigation headers, in the next section, you'll learn about tab and drawer navigation.

Tab and drawer navigation

So far in this chapter, each example has used `Button` components to link to other screens in the app. You can use functions from `react-navigation` that will create tab or drawer navigation for you automatically based on the screen components that you give it.

Let's create an example that uses bottom tab navigation on iOS and drawer navigation on Android.

> **Important Note**
> You aren't limited to using tab navigation on iOS or drawer navigation on Android. I'm just picking these two to demonstrate how to use different modes of navigation based on the platform. You can use the exact same navigation mode on both platforms if you prefer.

For this example, we need to install a few other packages for tab and drawer navigators:

```
npm install @react-navigation/bottom-tabs @react-navigation/
drawer
```

Also, the drawer navigator requires some native modules. Let's install them:

```
expo install react-native-gesture-handler react-native-
reanimated
```

Then, add a plugin to the `babel.config.js` file. As a result, the file should look like the following:

```
module.exports = function (api) {
  api.cache(true);
  return {
    presets: ["babel-preset-expo"],
    plugins: ["react-native-reanimated/plugin"],
  };
};
```

Now, we are ready to continue coding. Here's what the `App` component looks like:

```
import { NavigationContainer } from
  "@react-navigation/native";
import { createDrawerNavigator } from
  "@react-navigation/drawer";
```

```
import { createBottomTabNavigator } from
  "@react-navigation/bottom-tabs";
import { Platform } from "react-native";
import Home from "./Home";
import News from "./News";
import Settings from "./Settings";

const Tab = createBottomTabNavigator();
const Drawer = createDrawerNavigator();

export default function App() {
  return (
    <NavigationContainer>
      {Platform.OS === "ios" && (
        <Tab.Navigator>
          <Tab.Screen name="Home" component={Home} />
          <Tab.Screen name="News" component={News} />
          <Tab.Screen name="Settings" component={Settings}
            />
        </Tab.Navigator>
      )}

      {Platform.OS == "android" && (
        <Drawer.Navigator>
          <Drawer.Screen name="Home" component={Home} />
          <Drawer.Screen name="News" component={News} />
          <Drawer.Screen name="Settings"
            component={Settings} />
        </Drawer.Navigator>
      )}
    </NavigationContainer>
  );
}
```

Instead of using the `createNativeStackNavigator()` function to create your navigator, you're importing the `createBottomTabNavigator()` and `createDrawerNavigator()` functions:

```
import { createDrawerNavigator } from
  "@react-navigation/drawer";
import { createBottomTabNavigator } from
  "@react-navigation/bottom-tabs";
```

Then, you're using the `Platform` utility from `react-native` to decide which navigator to use. The result, depending on the platform, is assigned to `App`. Each navigator contains the `Navigator` and `Screen` components, and you can pass them to your `App`. The resulting tab or drawer navigation will be created and rendered for you:

```
export default function App() {
  return (
    <NavigationContainer>
      {Platform.OS === "ios" && (
        <Tab.Navigator>
          <Tab.Screen name="Home" component={Home} />
          <Tab.Screen name="News" component={News} />
          <Tab.Screen name="Settings" component={Settings}
            />
        </Tab.Navigator>
      )}

      {Platform.OS == "android" && (
        <Drawer.Navigator>
          <Drawer.Screen name="Home" component={Home} />
          <Drawer.Screen name="News" component={News} />
          <Drawer.Screen name="Settings"
            component={Settings} />
        </Drawer.Navigator>
      )}
    </NavigationContainer>
  );
}
```

Next, let's take a look at the Home screen component:

```
import React from "react";
import { View, Text } from "react-native";
import styles from "./styles";

export default function Home() {
  return (
    <View style={styles.container}>
      <Text>Home Content</Text>
    </View>
  );
}
```

The News and Settings components are essentially the same as Home. Here's what the bottom tab navigation looks like on iOS:

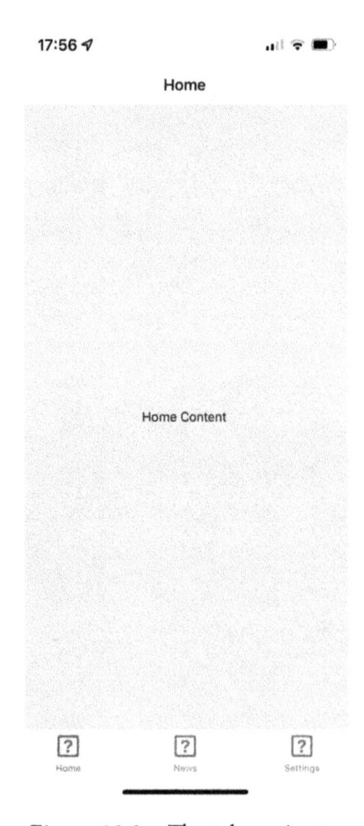

Figure 18.8 – The tab navigator

The three screens that make up your app are listed at the bottom. The current screen is marked as active, and you can click on the other tabs to move around.

Now, let's see what the drawer layout looks like on Android:

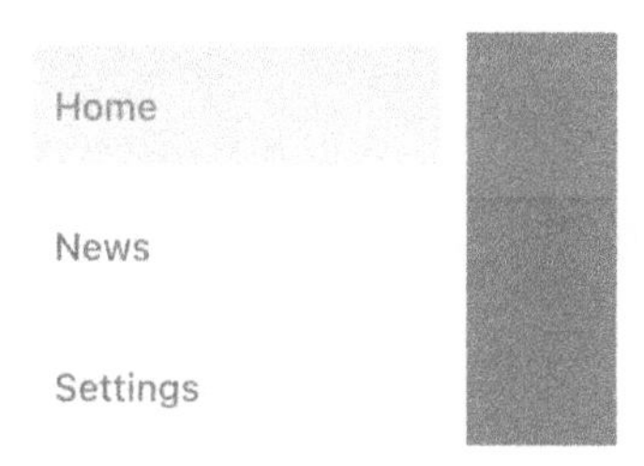

Figure 18.9 – The drawer navigator

To open the drawer, you need to swipe from the left side of the screen. Once it's open, you'll see buttons that will take you to the various screens of your app.

> **Important Note**
>
> Swiping the drawer open from the left side of the screen is default mode. You can configure the drawer to swipe open from any direction.

Now, you've learned how to use tab and drawer navigation.

Summary

In this chapter, you learned that mobile applications require navigation, just like web applications do. Although different, web application and mobile application navigation have enough conceptual similarities that mobile app routing and navigation don't have to be a nuisance.

Older versions of React Native made attempts to provide components to help manage navigation within mobile apps, but these never really took hold. Instead, the React Native community has dominated this area. One example of this is the `react-navigation` library, the focus of this chapter.

You learned how basic navigation works with `react-navigation`. You then learned how to control header components within the navigation bar. Next, you learned about tab and drawer navigation. These two navigation components can automatically render the navigation buttons for your app based on the screen components.

In the next chapter, you'll learn how to render lists of data.

Further reading

Check out the following link for more information on React Navigation: `https://reactnavigation.org/`.

19
Rendering Item Lists

In this chapter, you'll learn how to work with item lists. Lists are a common web application component. While it's relatively straightforward to build lists using the `<ul>` and `<li>` elements, doing something similar on native mobile platforms is much more involved.

Thankfully, React Native provides an item list interface that hides all of the complexity. First, you'll get a feel for how item lists work by walking through an example. Then, you'll learn how to build controls that change the data displayed in lists. Lastly, you'll see a couple of examples that fetch items from the network. The following are the sections you'll find in this chapter:

- Rendering data collections
- Sorting and filtering lists
- Fetching list data
- Lazy list loading
- Implementing pull to refresh

Technical requirements

You can find the code files for this chapter on GitHub at `https://github.com/PacktPublishing/React-and-React-Native-4th-Edition/tree/main/Chapter19`.

Rendering data collections

Lists are the most common way to display a lot of information – for example, you can display your friend list, messages, and news. Many apps contain lists with data collections, and React Native provides the tools to create these components.

Let's start with an example. The React Native component you'll use to render lists is FlatList, which works the same way on iOS and Android. List views accept a data property, which is an array of objects. These objects can have any properties you like, but they do require a key property. If you don't have a key property, you can pass the keyExtractor prop to the Flatlist component and instruct what to use instead of key. The key property is similar to the requirement for rendering the <li> elements inside of a <ul> element. This helps the list to efficiently render when changes are made to list data.

Let's implement a basic list now. Here's the code to render a basic 100-item list:

```
import React from "react";
import { Text, View, FlatList } from "react-native";
import styles from "./styles";

const data = new Array(100)
  .fill(null)
  .map((v, i) => ({ key: i.toString(), value: 'Item ${i}'
    }));

export default function App() {
  return (
    <View style={styles.container}>
      <FlatList
        data={data}
        renderItem={({ item }) => <Text
          style={styles.item}>{item.value}</Text>}
      />
    </View>
  );
}
```

Let's walk through what's going on here, starting with the `data` constant. This has an array of `100` items in it. It is created by filling a new array with `100` null values and then mapping this to a new array with the objects that you want to pass to `<FlatList>`. Each object has a `key` property because this is a requirement; anything else is optional. In this case, you've decided to add a `value` property that will be used later on when the list is rendered.

Next, you render the `<FlatList>` component. It's within a `<View>` container because list views need height in order to make scrolling work correctly. The `data` and `renderItem` properties are passed to `<FlatList>`, which ultimately determines the rendered content.

At first glance, it would seem that the `FlatList` component doesn't do too much. Do you have to figure out how the items look? Well, yes, the `FlatList` component is supposed to be generic. It's supposed to excel at handling updates and embeds scrolling capabilities into lists for us. Here are the styles that were used to render the list:

```
import { StyleSheet } from "react-native";

export default StyleSheet.create({
  container: {
    flex: 1,
    flexDirection: "column",
    paddingTop: 40,
  },

  item: {
    margin: 5,
    padding: 5,
    color: "slategrey",
    backgroundColor: "ghostwhite",
    textAlign: "center",
  },
});
```

Here, you're styling each item in your list. Otherwise, each item would be text-only, and it would be difficult to differentiate between other list items. The `container` style gives the list height by setting `flex` to `1`. Without height, you won't be able to scroll properly.

Let's see what the list looks like now:

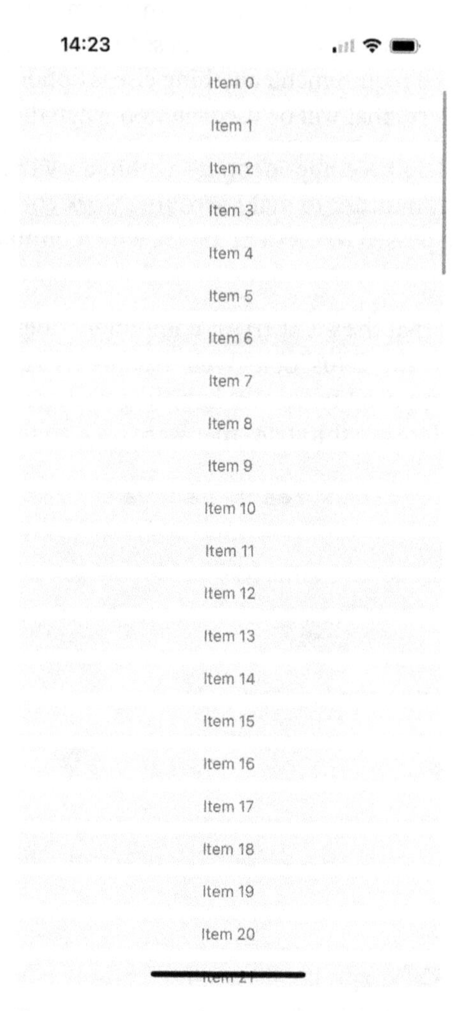

Figure 19.1 – Rendering the data collection

If you're running this example in a simulator, you can click and hold down the mouse button anywhere on the screen, like a finger, and then scroll up and down through the items.

In the following section, you'll learn how to add controls for sorting and filtering lists.

Sorting and filtering lists

Now that you have learned the basics of the `FlatList` components, including how to pass data, let's add some controls to the list that you just implemented in the *Rendering data collections* section. The `FlatList` component can be rendered together with other components – for example, list controls. It helps you to manipulate the data source, which ultimately drives what's rendered on the screen.

Before implementing list control components, it might be helpful to review the high-level structure of these components so that the code has more context. Here's an illustration of the component structure that you're going to implement:

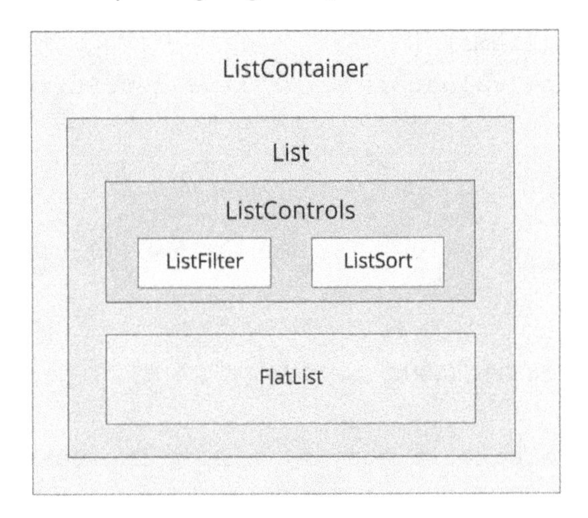

Figure 19.2 – The component structure

Here's what each of these components is responsible for:

- `ListContainer`: The overall container for the list; it follows the familiar React container pattern
- `List`: A stateless component that passes the relevant pieces of state into `ListControls` and the React Native `ListView` component
- `ListControls`: A component that holds the various controls that change the state of the list
- `ListFilter`: A control for filtering the item list
- `ListSort`: A control for changing the sort order of the list
- `FlatList`: The actual React Native component that renders items

In some cases, splitting apart the implementation of a list like this is overhead. However, I think that if your list needs controls in the first place, you're probably implementing something that will stand to benefit from having a well-thought-out component architecture.

Now, let's drill down into the implementation of this list, starting with the `ListContainer` component:

```
import React, { useState, useMemo } from "react";
import List from "./List";

function mapItems(items) {
  return items.map((value, i) => ({ key: i.toString(),
    value }));
}

const array = new Array(100).fill(null).map((v, i) =>
  'Item ${i}');

function filterAndSort(text, asc) {
  return array
    .filter((i) => text.length === 0 || i.includes(text))
    .sort(
      asc
        ? (a, b) => (a > b ? 1 : a < b ? -1 : 0)
        : (a, b) => (b > a ? 1 : b < a ? -1 : 0)
    );
}
```

Here, we define a few utility functions and the initial array that we will use.

Then, we will define `asc` and `filter` for managing sorting and filtering the list, respectively, with the `data` variable implemented using the `useMemo` hook:

```
export default function ListContainer() {
  const [asc, setAsc] = useState(true);
  const [filter, setFilter] = useState("");
```

```
  const data = useMemo(() => {
    return filterAndSort(filter, asc);
  }, [filter, asc]);
```

It gives us an opportunity to avoid updating it manually because it will be recalculated automatically when the `filter` and `asc` dependencies are updated. It also helps us to avoid unnecessary recalculation when `filter` and `asc` are not changed.

Here is how we apply this logic to the `List` component:

```
  return (
    <List
      data={mapItems(data)}
      asc={asc}
      onFilter={(text) => {
        setFilter(text);
      }}
      onSort={() => {
        setAsc(!asc);
      }}
    />
  );
}
```

If this seems like a bit much, it's because it is. This container component has a lot of state to handle. It also has some non-trivial behavior that it needs to make available to its children. If you look at it from the perspective of an encapsulating state, it will be more approachable. Its job is to populate the list with state data and provide functions that operate in this state.

In an ideal world, the child components of this container should be nice and simple, since they don't have to directly interface with the state. Let's take a look at the `List` component next:

```
import React from "react";
import PropTypes from "prop-types";
import { Text, FlatList } from "react-native";
import styles from "./styles";
```

```
import ListControls from "./ListControls";

export default function List({ Controls, data, onFilter,
  onSort, asc }) {
  return (
    <FlatList
      data={data}
      ListHeaderComponent={<Controls {...{ onFilter,
        onSort, asc }} />}
      renderItem={({ item }) => <Text
        style={styles.item}>{item.value}</Text>}
    />
  );
}

List.propTypes = {
  Controls: PropTypes.func.isRequired,
  data: PropTypes.array.isRequired,
  onFilter: PropTypes.func.isRequired,
  onSort: PropTypes.func.isRequired,
  asc: PropTypes.bool.isRequired,
};

List.defaultProps = {
  Controls: ListControls,
};
```

This component takes the state from the `ListContainer` component as properties and renders a `FlatList` component. The main difference here from the previous example is the `ListHeaderComponent` property. This renders the controls for your `List`. What's especially useful about this property is that it renders the controls outside the scrollable list content, ensuring that the controls are always visible.

Also, note that you're specifying your own `ListControls` component as a default value for the `Controls` property. This makes it easy for others to pass in their own list controls. Let's take a look at the `ListControls` component next:

```
import React from "react";
import PropTypes from "prop-types";
import { View } from "react-native";
import styles from "./styles";
import ListFilter from "./ListFilter";
import ListSort from "./ListSort";

export default function ListControls({ onFilter, onSort,
  asc }) {
  return (
    <View style={styles.controls}>
      <ListFilter onFilter={onFilter} />
      <ListSort onSort={onSort} asc={asc} />
    </View>
  );
}

ListControls.propTypes = {
  onFilter: PropTypes.func.isRequired,
  onSort: PropTypes.func.isRequired,
  asc: PropTypes.bool.isRequired,
};
```

This component brings together the `ListFilter` and `ListSort` controls. So, if you were to add another list control, you would add it here.

Let's take a look at the `ListFilter` implementation now:

```
import React from "react";
import PropTypes from "prop-types";
import { View, TextInput } from "react-native";
import styles from "./styles";
```

```
export default function ListFilter({ onFilter }) {
  return (
    <View>
      <TextInput
        autoFocus
        placeholder="Search"
        style={styles.filter}
        onChangeText={onFilter}
      />
    </View>
  );
}

ListFilter.propTypes = {
  onFilter: PropTypes.func.isRequired,
};
```

The filter control is a simple text input that filters the list of items by user type. The `onChange` function that handles this comes from the `ListContainer` component.

Let's look at the `ListSort` component next:

```
import React from "react";
import PropTypes from "prop-types";
import { Text } from "react-native";

const arrows = new Map([
  [true, "▼"],
  [false, "▲"],
]);

export default function ListSort({ onSort, asc }) {
  return <Text onPress={onSort}>{arrows.get(asc)}</Text>;
}
```

```
ListSort.propTypes = {
  onSort: PropTypes.func.isRequired,
  asc: PropTypes.bool.isRequired,
};
```

Here's a look at the resulting list:

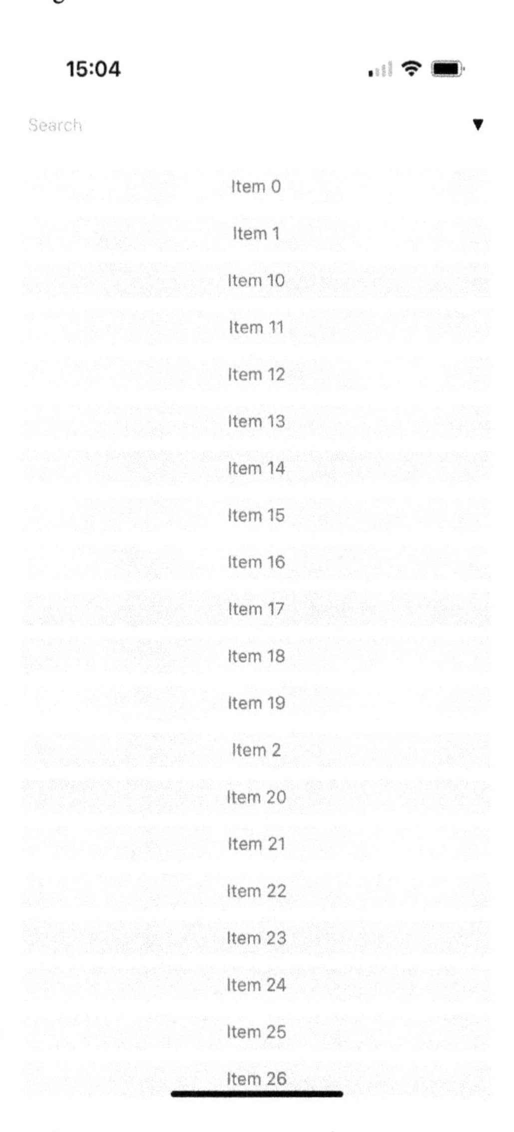

Figure 19.3 – The sorting and filtering list

By default, the entire list is rendered in ascending order. You can see the placeholder **Search** text when the user hasn't provided anything yet. Let's see how this looks when you enter a filter and change the sort order:

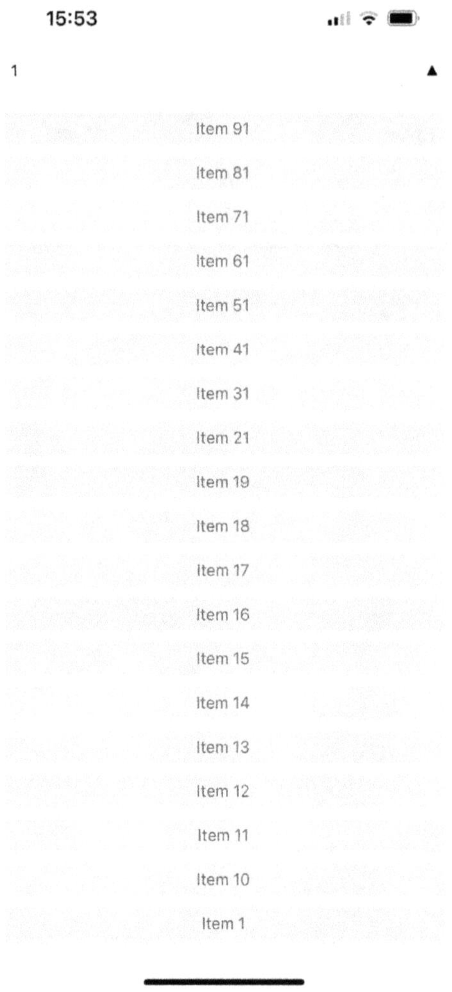

Figure 19.4 – The list with a changed sort order and search value

This search includes items containing **1** and sorts the results in descending order. Note that you can either change the order first or enter the filter first. Both the filter and the sort order are part of the `ListContainer` state.

In the next section, you'll learn how to fetch list data from an API endpoint.

Fetching list data

Often, you'll fetch your list data from some API endpoint. In this section, you'll learn about making API requests from React Native components. The good news is that the `fetch()` API is polyfilled by React Native, so the networking code in your mobile applications should look and feel a lot like it does in your web applications.

To start things off, let's build a mock API for our list items, using functions that return promises just like `fetch()` does:

```
const items = new Array(100).fill(null).map((v, i) =>
  'Item ${i}');

function filterAndSort(data, text, asc) {
  return data
    .filter(i => text.length === 0 || i.includes(text))
    .sort(
      asc
        ? (a, b) => (b > a ? -1 : a === b ? 0 : 1)
        : (a, b) => (a > b ? -1 : a === b ? 0 : 1)
    );
}

export function fetchItems(filter, asc) {
  return new Promise(resolve => {
    resolve({
      json: () =>
        Promise.resolve({
          items: filterAndSort(items, filter, asc)
        })
    });
  });
}
```

With the mock API function in place, let's make some changes to the `ListContainer` component. Instead of using local data sources, you can now use the `fetchItems()` function to load data from the mock API:

```
import React, { useState, useEffect } from "react";
import { fetchItems } from "./api";
import List from "./List";

function mapItems(items) {
  return items.map((value, i) => ({ key: i.toString(),
    value }));
}
```

In the preceding block, we've imported the necessary components and methods and defined the `mapItems` function to avoid repetition of such logic.

Let's take a look and define the `ListContainer` component:

```
export default function ListContainer() {
  const [asc, setAsc] = useState(true);
  const [filter, setFilter] = useState("");
  const [data, setData] = useState([]);

  useEffect(() => {
    fetchItems(filter, asc)
      .then((resp) => resp.json())
      .then(({ items }) => {
        setData(mapItems(items));
      });
  }, []);
```

We've defined state variables using the `useState` and `useEffect` Hooks to fetch initial list data.

Now, let's take a look at the usage of our new handlers in the `List` component:

```
  return (
    <List
      data={data}
      asc={asc}
      onFilter={(text) => {
        fetchItems(text, asc)
          .then((resp) => resp.json())
          .then(({ items }) => {
            setFilter(text);
            setData(mapItems(items));
          });
      }}
      onSort={() => {
        fetchItems(filter, !asc)
          .then((resp) => resp.json())
          .then(({ items }) => {
            setAsc(!asc);
            setData(mapItems(items));
          });
      }}
    />
  );
}
```

Any action that modifies the state of the list needs to call `fetchItems()` and set the appropriate state once the promise resolves.

In the following section, you'll learn how list data can be loaded lazily.

Lazy list loading

In this section, you'll implement a different kind of list – one that scrolls infinitely. Sometimes, users don't actually know what they're looking for, so filtering or sorting isn't going to help. Think about the Facebook news feed you see when you log in to your account; it's the main feature of the application, and rarely are you looking for something specific. You need to see what's going on by scrolling through the list.

To do this using a `FlatList` component, you need to be able to fetch more API data when the user scrolls to the end of the list. To get an idea of how this works, you need a lot of API data to work with, and generators are great at this. So, let's modify the mock that you created in the *Fetching list data* section's example so that it just keeps responding with new data:

```
function* genItems() {
  let cnt = 0;

  while (true) {
    yield 'Item ${cnt++}';
  }
}

const items = genItems();

export function fetchItems() {
  return Promise.resolve({
    json: () =>
      Promise.resolve({
        items: new Array(20).fill(null).map(() =>
          items.next().value),
      }),
  });
}
```

With `fetchItems`, you can now make an API request for new data every time the end of the list is reached. Eventually, this will fail when you run out of memory, but I'm just trying to show you in general terms the approach you can take to implement infinite scrolling in React Native. Now, let's take a look at what the `ListContainer` component looks like with `fetchItems`:

```
import React, { useState, useEffect } from "react";
import * as api from "./api";
import List from "./List";

export default function ListContainer() {
  const [data, setData] = useState([]);

  function fetchItems() {
    return api
      .fetchItems()
      .then((resp) => resp.json())
      .then(({ items }) => {
        setData([
          ...data,
          ...items.map((value) => ({
            key: value,
            value,
          })),
        ]);
      });
  }

  useEffect(() => {
    fetchItems();
  }, []);

  return <List data={data} fetchItems={fetchItems} />;
}
```

Each time `fetchItems()` is called, the response is concatenated with the `data` array. This becomes the new list data source, instead of replacing it as you did in earlier examples.

Now, let's take a look at the `List` component to see how to respond to the end of the list being reached:

```
import React from "react";
import PropTypes from "prop-types";
import { Text, FlatList } from "react-native";
import styles from "./styles";

export default function List({ data, fetchItems }) {
  return (
    <FlatList
      data={data}
      renderItem={({ item }) => <Text
        style={styles.item}>{item.value}</Text>}
      onEndReached={fetchItems}
    />
  );
}

List.propTypes = {
  data: PropTypes.array.isRequired,
  fetchItems: PropTypes.func.isRequired,
};
```

`FlatList` accepts the `onEndReached` handler prop, which will be invoked every time you reach the end of the list during scrolling.

If you run this example, you'll see that, as you approach the bottom of the screen while scrolling, the list just keeps growing.

Implementing pull to refresh

The pull to refresh gesture is a common action on mobile devices. It allows users to refresh the content of a view without having to lift a finger from the screen or manually reopen the app, just by pulling it down to trigger a page refresh. Loren Brichter, the creator of Tweetie (later Twitter for iPhone) and Letterpress, introduced this gesture in 2009. This gesture has become so popular that Apple integrated it into its SDKs as `UIRefreshControl`.

To use pull to refresh in the `FlatList` app, we just need to pass a few props and handlers. Let's take a look at our `List` component:

```
export default function List({ data, fetchItems,
  refreshItems, isRefreshing }) {
  return (
    <FlatList
      data={data}
      renderItem={({ item }) => <Text
        style={styles.item}>{item.value}</Text>}
      onEndReached={fetchItems}
      onRefresh={refreshItems}
      refreshing={isRefreshing}
    />
  );
}
```

As we have provided the `onRefresh` and `refreshing` props, our `FlatList` component automatically enables the pull to refresh gesture. The `onRefresh` handler will be called when you pull the list, and the `refreshing` property will enable the loading spinner to reflect the loading state.

To apply defined props in the `List` component, let's implement the `refreshItems` function with the `isRefreshing` state in the `ListContainer` component:

```
const [isRefreshing, setIsRefreshing] = useState(false);
function refreshItems() {
    setIsRefreshing(true);
    return api
      .fetchItems({ refresh: true })
      .then((resp) => resp.json())
      .then(({ items }) => {
```

```
        setData(
          items.map((value) => ({
            key: value,
            value,
          }))
        );
      })
      .finally(() => {
        setIsRefreshing(false);
      });
  }
```

In `refreshItems`, as well as in the `fetchItems` method, we get list items but save them as a new list. Also, note that before calling the API, we update the `isRefreshing` state to set it as a `true` value, and in the `finally` block, we set it to `false` to provide information to `FlatList` that loading has ended.

Summary

In this chapter, you learned about the `FlatList` component in React Native. This component is general-purpose, as it doesn't impose any specific look for items that get rendered. Instead, the appearance of the list is up to you, leaving the `FlatList` component to help with efficiently rendering a data source. The `FlatList` component also provides a scrollable region for the items it renders.

You implemented an example that took advantage of section headers in list views. This is a good place to render static content such as list controls. You then learned about making network calls in React Native; it's just like using `fetch()` in any other web application.

Finally, you implemented lazy lists that scroll infinitely by only loading new items after they've scrolled to the bottom of what's already been rendered. Also, we added a feature to refresh that list by means of a pull gesture.

In the next chapter, you'll learn how to show the progress of network calls, among other things.

Further reading

Take a look at the following link for more information on `FlatList`: `https://reactnative.dev/docs/flatlist`.

20
Showing Progress

This chapter is all about communicating progress to the user. React Native has different components that are used to handle the different types of progress that you want to communicate. First, you'll learn why you need to communicate progress in the app. Then, you'll learn how to implement progress indicators and progress bars. And finally, you'll see specific examples that show you how to use progress indicators with navigation while data loads and progress bars to communicate the current position in a series of steps.

The following sections are covered in this chapter:

- Understanding progress and usability
- Indicating progress
- Measuring progress
- Exploring navigation indicators
- Step progress

Technical requirements

You can find the code files for this chapter on GitHub at `https://github.com/PacktPublishing/React-and-React-Native-4th-Edition/tree/main/Chapter20`.

Understanding progress and usability

Imagine that you have a microwave oven that has no window and makes no sound. The only way to interact with it is by pressing a button labeled *cook*. As absurd as this device sounds, it's what many software users face – no indication of progress. Is the microwave cooking anything? If so, how do we know when it will be done?

One way to improve the microwave situation is to add sound. This way, the user gets feedback after pressing the *cook* button. You've overcome one hurdle, but the user is still left asking, "Where's my food?" Before you go out of business, you had better add some sort of progress measurement display, such as a timer.

It's not that UI programmers don't understand the basic principles of this usability concern; it's just that they have stuff to do, and this sort of thing simply slips through the cracks in terms of priority. In React Native, there are components to give the user indeterminate progress feedback and precise progress measurements. It's always a good idea to make these things a top priority if you want a good user experience.

Now that you understand the role of progress in usability, it's time to learn how to indicate progress in your React Native UIs.

Indicating progress

In this section, you'll learn how to use the `ActivityIndicator` component. As its name suggests, you render this component when you need to indicate to the user that something is happening. The actual progress may be indeterminate, but at least you have a standardized way to show that something is happening, despite there being no results to display yet.

Let's create an example so that you can see what this component looks like. Here's the `App` component:

```
import React from "react";
import { View, ActivityIndicator } from "react-native";
import styles from "./styles";

export default function App() {
  return (
    <View style={styles.container}>
      <ActivityIndicator size="large" />
    </View>
```

```
    );
}
```

The <ActivityIndicator /> component is platform-agnostic. Here's how it looks on iOS:

Figure 20.1 – An activity indicator on iOS

It renders an animated spinner in the middle of the screen. This is the large spinner, as specified in the `size` property. The `ActivityIndicator` spinner can also be small, which makes more sense if you're rendering it inside another smaller element. Now, let's take a look at how this looks on an Android device:

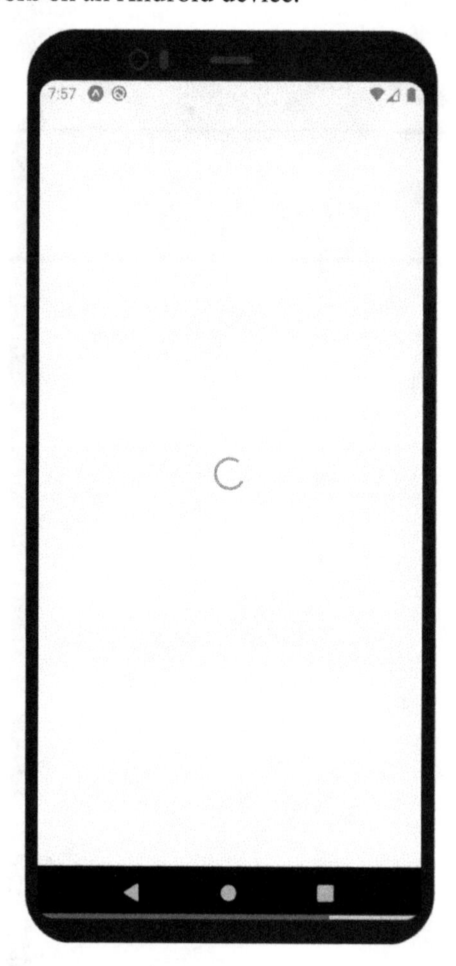

Figure 20.2 – An activity indicator on Android

The spinner looks different, as it should, but your app conveys the same thing on both platforms – you're waiting for something.

This example spins forever. But don't worry – there's a more realistic progress indicator example coming up that shows you how to work with navigation and loading API data.

Measuring progress

The downside of just indicating that progress is being made is that there's no end in sight for the user. This leads to a feeling of unease, like when you're waiting for food to cook in a microwave with no timer. When you know how much progress has been made and how much is left to go, you feel better. That is why it's always better to use a deterministic progress bar whenever possible.

Unlike the `ActivityIndicator` component, there's no platform-agnostic component in React Native for progress bars. So, we'll have to make one ourselves. We'll create a component that uses `ProgressViewIOS` on iOS and `ProgressBarAndroid` on Android.

> **Important Information**
>
> Due to `react-native` size optimization, the Meta team is working on moving such components to separate packages. In the next releases, `ProgressViewIOS` and `ProgressBarAndroid` might be moved outside of the react-native library.
>
> You can also try the following packages with a similar API:
>
> **expo-progress** – `https://github.com/EvanBacon/expo-progress` and
>
> **ProgressBar** – `https://callstack.github.io/react-native-paper/progress-bar.html`.

Let's handle the cross-platform issues first. React Native knows to import the correct module based on its file extension. Here's what the `ProgressBarComponent.ios.js` module looks like:

```
export { ProgressViewIOS as ProgressBarComponent } from
  "react-native";
export const progressProps = {};
```

You're directly exporting the `ProgressViewIOS` component from React Native. You're also exporting properties for a component specific to the platform. In this case, it's an empty object because there are no properties that are specific to `<ProgressViewIOS>`.

Now, let's look at the `ProgressBarComponent.android.js` module:

```
export { ProgressBarAndroid as ProgressBarComponent } from
  "react-native";

export const progressProps = {
  styleAttr: "Horizontal",
  indeterminate: false
};
```

This module uses the exact same approach as the `ProgressBarComponent.ios.js` module. It exports the Android-specific component and the Android-specific properties to pass to it.

Now, let's build the `ProgressBar` component that the application will use. You can find the source code for this component here: `https://github.com/PacktPublishing/React-and-React-Native-4th-Edition-/blob/main/Chapter20/measuring-progress/ProgressBar.js`.

Let's walk through what's going on in this module, starting with the imports. The `ProgressBarComponent` and `progressProps` values are imported from our `ProgressBarComponent` module. React Native determines which module to import these from.

Next, you have the `ProgressLabel` utility component. It figures out what label is rendered for the progress bar based on the `show` property. If it's `false`, nothing is rendered. If it's `true`, it renders a `<Text>` component that displays the progress as a percentage.

Lastly, you have the `ProgressBar` component itself, which our application will import and use. This renders the label and the appropriate progress bar component. It takes a `progress` property, which is a value between `0` and `1`. Now, let's put this component to use in the `App` component:

```
import React, { useState, useEffect } from "react";
import { View } from "react-native";
import styles from "./styles";
import ProgressBar from "./ProgressBar";

export default function MeasuringProgress() {
  const [progress, setProgress] = useState(0);

  useEffect(() => {
    function updateProgress() {
      setProgress(currentProgress => {
        if (currentProgress < 1) {
          setTimeout(updateProgress, 300);
          return currentProgress + 0.01;
        }
        return currentProgress;
      });
    }

    updateProgress();
  }, []);

  return (
    <View style={styles.container}>
      <ProgressBar progress={progress} />
    </View>
  );
}
```

Initially, the `<ProgressBar>` component is rendered at 0%. In the `useEffect()` hook, the `updateProgress()` function uses a timer to simulate a real process that you want to show progress for. Here's what the iOS screen looks like:

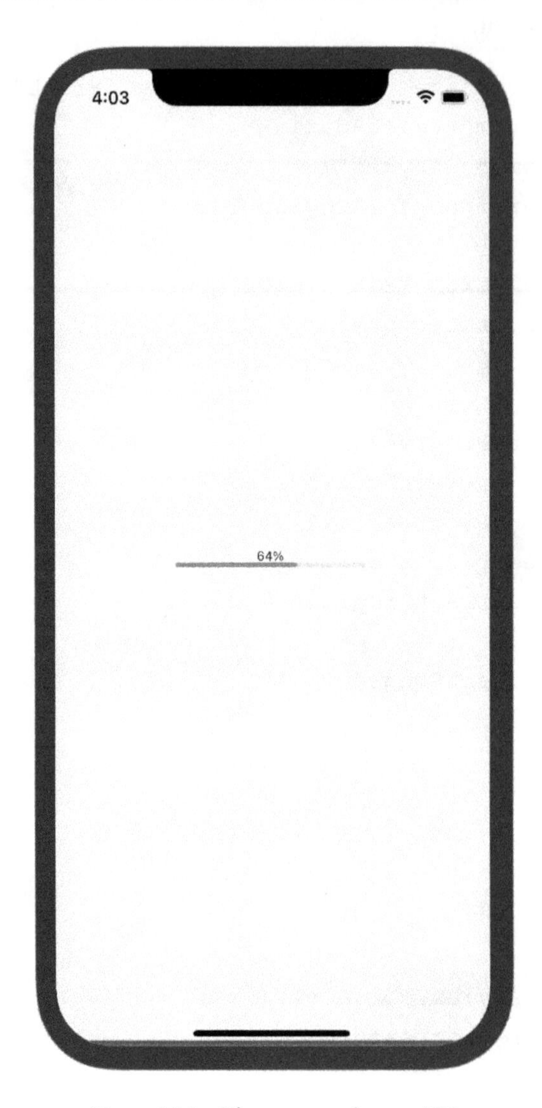

Figure 20.3 – The progress bar on iOS

Here's what the same progress bar looks like on Android:

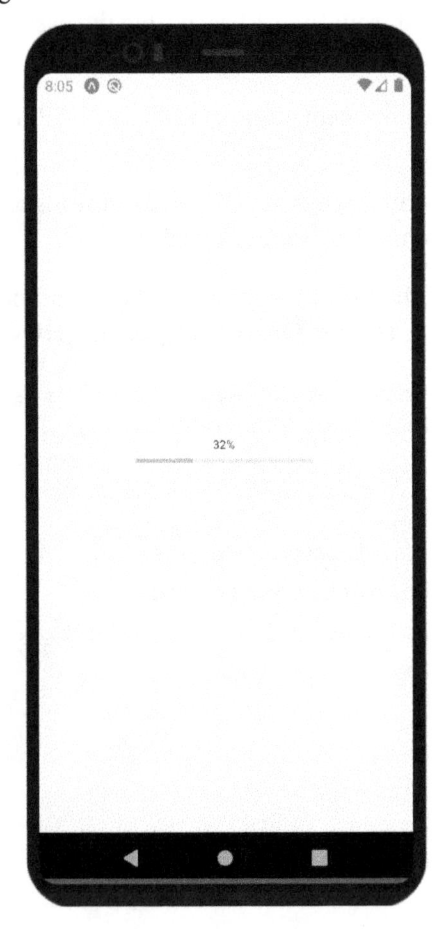

Figure 20.4 – The progress bar on Android

Showing a quantitative measure of progress is important so that users can gauge how long something will take. In the next section, you'll learn how to use progress indicators to show the user where they are in terms of navigating screens.

Exploring navigation indicators

Earlier in this chapter, you were introduced to the `ActivityIndicator` component. In this section, you'll learn how it can be used when navigating an application that loads data. For example, the user navigates from page or screen one to page two. However, page two needs to fetch data from the API that it can display to the user. So, while this network call is happening, it makes more sense to display a progress indicator instead of a screen devoid of useful information.

Doing this is actually kind of tricky because you have to make sure that the data that's required by the screen is fetched from the API each time the user navigates to the screen. Your goals should be as follows:

- Have the `Navigator` component automatically fetch API data for the scene that's about to be rendered.
- Use the promise that's returned by the API call as a means to display the spinner and hide it once the promise has been resolved.

Since your components probably don't care about whether a spinner is displayed or not, let's implement this as a generic **Higher-Order Component** (**HOC**):

```
import React, { useState, useEffect } from "react";
import { View, ActivityIndicator } from "react-native";
import styles from "./styles";

export default function loading(Wrapped) {
  return function LoadingWrapper(props) {
    const [loading, setLoading] = useState(true);

    useEffect(() => {
      setTimeout(() => {
        setLoading(false);
      }, 1000);
    }, []);

    if (loading) {
      return (
        <View style={styles.container}>
          <ActivityIndicator size="large" />
        </View>
      );
    } else {
      return <Wrapped {...props} />;
    }
  };
}
```

This `loading()` function takes a component – the `Wrapped` argument – and returns a `LoadingWrapper` component. The HOC has a `useEffect` hook with a `timeout`, and when it resolves, it changes the `loading` state to `false`. As you can see in the `render()` method, the `loading` state determines whether the spinner or the `Wrapped` component is rendered.

With the `loading()` higher-order function in place, let's take a look at the first `screen` component that you'll use with `react-navigation`:

```
import React from "react";
import { View, Button } from "react-native";
import styles from "./styles";
import loading from "./loading";

const First = loading(({ navigation }) => (
  <View style={styles.container}>
    <Button title="Second" onPress={() =>
      navigation.navigate("Second")} />
    <Button title="Third" onPress={() =>
      navigation.navigate("Third")} />
  </View>
));

export default First;
```

This module exports a component that's wrapped with the `loading()` HOC we created earlier. It wraps the `First` component so that a spinner is displayed while the `setTimeout` method is pending. This is a useful approach to hiding extra logic in one place and reusing it on every page. Instead of the `setTimeout` method, in a real app, you can pass additional props to the HOC.

Step progress

In this final example, you'll build an app that displays the user's progress through a predefined number of steps. For example, it might make sense to split a form into several logical sections and organize them in such a way that, as the user completes one section, they move to the next step. A progress bar would be helpful feedback for the user.

You'll insert a progress bar into the navigation bar, just below the title, so that the user knows how far they've gone and how far is left to go. You'll also reuse the `ProgressBar` component that you implemented earlier in this chapter.

Let's take a look at the result first. There are four screens in this app that the user can navigate. Here's what the first page (scene) looks like:

Figure 20.5 – The first screen

The progress bar under the title reflects the fact that the user is 25% through the navigation. Let's see what the third screen looks like:

Figure 20.6 – The third screen

The progress is updated to reflect where the user is in the route stack. Let's take a look at the `App` component here: `https://github.com/PacktPublishing/React-and-React-Native-4th-Edition-/blob/main/Chapter20/step-progress-new/App.js`.

This app has four screens. The components that render each of these screens are stored in the `routes` constant, which is then used to configure the stack navigator using `createNativeStackNavigator()`. The reason for creating the `routes` array is so that it can be used by the `progress` parameter that is passed by `initialParams` to every route. To calculate the progress, we take the current route index as a value of the route's length. For example, `Second` is in the number `2` position (an index of 1 + 1) and the length of the array is `4`. This will set the progress bar to 50%.

Also, the `Next` and `Previous` buttons' calls to `navigation.navigate()` have to pass `routeName`, so we added the `nextRouteName` and `prevRouteName` variables to the `screenOptions` handler.

Summary

In this chapter, you learned how to show your users that something is happening behind the scenes. First, we discussed why showing progress is important for the usability of an application. Then, we implemented a basic screen that indicated progress was being made. After that, we implemented a `ProgressBar` component, which is used to measure specific progress amounts.

Indicators are good for indeterminate progress. We implemented navigation that showed progress indicators while network calls were pending. In the final section, we implemented a progress bar that showed the user where they were in a predefined number of steps.

In the next chapter, we'll look at React Native maps and geolocation data in action.

Further reading

Check out the following links for more information:

- `ActivityIndicator`: `https://reactnative.dev/docs/activityindicator`
- `ProgressViewIOS`: `https://reactnative.dev/docs/progressviewios`
- `ProgressBarAndroid`: `https://reactnative.dev/docs/progressbarandroid`

21
Geolocation and Maps

In this chapter, you'll learn about the geolocation and mapping capabilities of React Native. You'll start the learning process with how to use the geolocation API, and then you'll move on to using the `MapView` component to plot points of interest and regions. To do this, we'll use the `react-native-maps` package to implement maps.

The goal of this chapter is to go over what's available in React Native for geolocation and in React Native Maps for maps.

Here's a list of the topics that we'll cover in this chapter:

- Using Location API
- Rendering the Map
- Annotating points of interest

Technical requirements

You can find the code file for this chapter on GitHub at `https://github.com/PacktPublishing/React-and-React-Native-4th-Edition/tree/main/Chapter21`.

Using Location API

The geolocation API that web applications use to figure out where the user is located can also be used by React Native applications because the same API has been polyfilled. Other than maps, this API is useful for getting precise coordinates from the GPS on mobile devices. You can then use this information to display meaningful location data to the user.

Unfortunately, the data returned by the geolocation API is of little use on its own. Your code must do the legwork to transform it into something useful. For example, latitude and longitude don't mean anything to the user, but you can use this data to look up something that is of use to the user. This might be as simple as displaying where the user is currently located.

Let's implement an example that uses the geolocation API of React Native to look up coordinates and then use those coordinates to look up human-readable location information from the Google Maps API.

Before we start coding, let's create a project using `expo init` and then add the location module:

```
expo install expo-location
```

When you have a prepared project, let's have a look at the `App` component, which you can find here: `https://github.com/PacktPublishing/React-and-React-Native-4th-Edition/blob/main/Chapter21/where-am-i/App.js`. The goal of this component is to render the properties returned by the geolocation API on the screen, as well as looking up the user's specific location and displaying it.

To fetch a location from the app, we need to grant permissions. In `App.js`, we have called `Location.requestForegroundPermissionsAsync()` for that.

The `setPosition()` function is used as a callback in a couple of places, with its job being to set the state of your component. Firstly, `setPosition()` sets the lat-long coordinates. Normally, you wouldn't display this data directly, but this is an example that's showing the data that's available as part of the geolocation API. And secondly, it uses the `latitude` and `longitude` values to look up the name of where the user is currently, using the Google Maps API.

In the example, the `API_KEY` value is empty, and you can get it here: `https://developers.google.com/maps/documentation/geocoding/start`.

The `setPosition()` callback is used with `getCurrentPosition()`, which is only called once when the component is mounted. You're also using `setPosition()` with `watchPosition()`, which calls the callback any time the user's position changes.

> **Important Note**
>
> The iOS emulator and Android Studio let you change locations via menu options. You don't have to install your app on a physical device every time you want to test changing locations.

Let's see what this screen looks like once the location data has loaded:

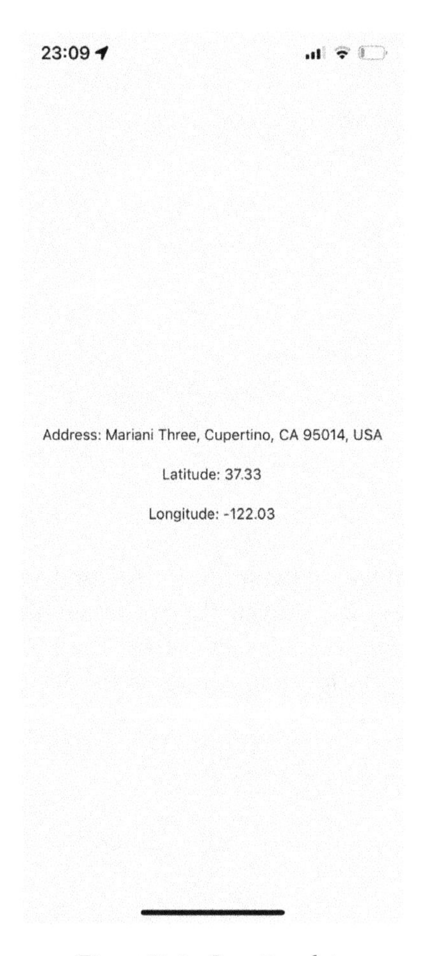

Figure 21.1 – Location data

The address information that was fetched is probably more useful in an application than latitude and longitude data. Even better than physical address text is visualizing the user's physical location on a map; you'll learn how to do this in the next section.

Rendering the Map

The `MapView` component from `react-native-maps` is the main tool you'll use to render maps in your React Native applications. It offers a wide range of tools for rendering maps, markers, polygons, heatmaps, and suchlike.

Let's now implement a basic `MapView` component to see what you get out of the box:

```
import React from "react";
import { View, StatusBar } from "react-native";
import MapView from "react-native-maps";
import styles from "./styles";

StatusBar.setBarStyle("dark-content");

export default () => (
  <View style={styles.container}>
    <MapView style={styles.mapView} showsUserLocation
      followUserLocation />
  </View>
);
```

The two Boolean properties that you've passed to `MapView` do a lot of work for you. The `showsUserLocation` property will activate the marker on the map, which denotes the physical location of the device running this application. The `followUserLocation` property tells the map to update the location marker as the device moves around.

Here is the resulting map:

Figure 21.2 – Current location

The current location of the device is clearly marked on the map. By default, points of interest are also rendered on the map. These are things close to the user so that they can see what's around them.

It's generally a good idea to use the `followUserLocation` property whenever using `showsUserLocation`. This makes the map zoom to the region where the user is located.

In the following section, you'll learn how to annotate points of interest on your maps.

Annotating points of interest

Annotations are exactly what they sound like; additional information rendered on top of the basic map geography. You get annotations by default when you render `MapView` components. The `MapView` component can render the user's current location and points of interest around the user. The challenge here is that you probably want to show the points of interest relevant to your application instead of those rendered by default.

In this section, you'll learn how to render markers for specific locations on the map, as well as rendering regions on the map.

Plotting points

Let's plot some local breweries! Here's how you pass annotations to the `MapView` component:

```
import React from "react";
import { View, StatusBar } from "react-native";
import MapView from "react-native-maps";
import styles from "./styles";

StatusBar.setBarStyle("dark-content");

export default function App() {
  return (
    <View style={styles.container}>
      <MapView
        style={styles.mapView}
        showsPointsOfInterest={false}
        showsUserLocation
        followUserLocation
      >
        <MapView.Marker
          title="Duff Brewery"
          description="Duff beer for me, Duff beer for you"
          coordinate={{
            latitude: 43.8418728,
            longitude: -79.086082,
          }}
```

```
        />
        <MapView.Marker
          title="Pawtucket Brewery"
          description="New! Patriot Light!"
          coordinate={{
            latitude: 43.8401328,
            longitude: -79.085407,
          }}
        />
      </MapView>
    </View>
  );
}
```

In this example, we've opted out of this capability by setting the `showsPointsOfInterest` property to `false`. Let's see where these breweries are located:

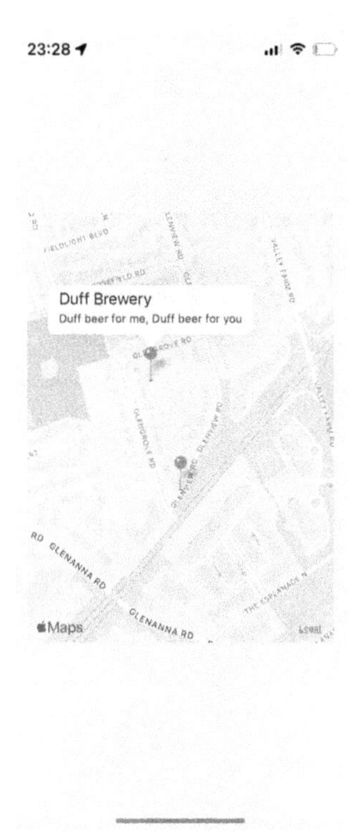

Figure 21.3 – Plotting points

The callout is displayed when you press the marker that shows the location of the brewery on the map. The `title` and `description` property values that you give to `<MapView.Marker>` are used to render this text.

Plotting overlays

In this last section of this chapter, you'll learn how to render region overlays. Think of a region as a connect-the-dots drawing of several points, and a point is a single `latitude/longitude` coordinate.

Regions can serve many purposes. In our example, we'll create a region that shows where we're more likely to find IPA drinkers versus stout drinkers. You can follow this link to see what the code looks like: `https://github.com/PacktPublishing/React-and-React-Native-4th-Edition/blob/main/Chapter21/plotting-overlays/App.js`.

The region data consists of several `latitude/longitude` coordinates that define the shape and location of the region. The rest of this code is mostly about the handling state when the two text links are pressed.

By default, the IPA region is rendered as follows:

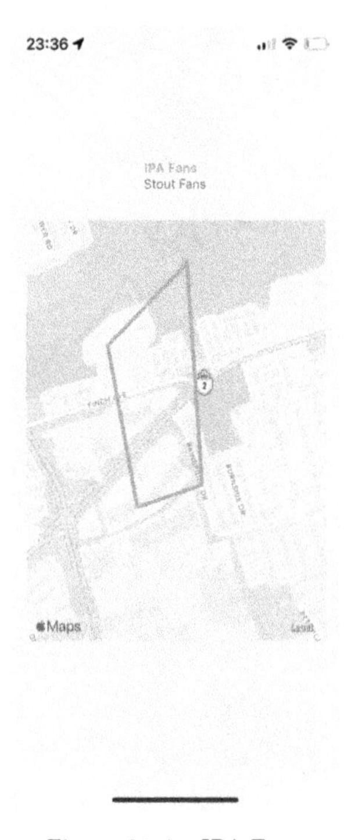

Figure 21.4 – IPA Fans

When the **Stout Fans** button is pressed, the IPA overlay is removed from the map, and the stout region is added:

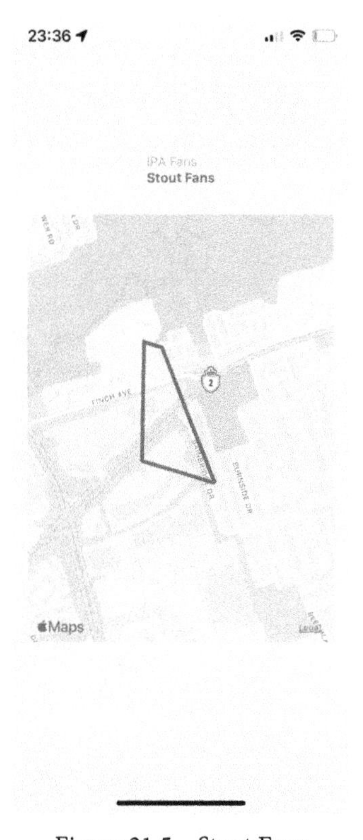

Figure 21.5 – Stout Fans

Overlays are useful when you need to highlight an area instead of a `latitude/longitude` point or an address.

Summary

In this chapter, you learned about geolocation and mapping in React Native. The geolocation API works the same as its web counterpart. The only reliable way to use maps in React Native applications is to install the third-party `react-native-maps` package.

You saw the basic configuration `MapView` components and how they can track the user's location and show relevant points of interest. Then, you saw how to plot your own points of interest and regions of interest.

In the next chapter, you'll learn how to collect user input using React Native components that resemble HTML form controls.

Further reading

Take a look at the following URLs to get more information:

- Geolocation: `https://docs.expo.dev/versions/latest/sdk/location/`
- React Native maps: `https://docs.expo.dev/versions/latest/sdk/map-view/`

22 Collecting User Input

In web applications, you can collect user input from standard HTML form elements that look and behave similarly on all browsers. With native UI platforms, collecting user input is more nuanced.

In this chapter, you'll learn how to work with the various React Native components that are used to collect user input. These include text input, selecting from a list of options, checkboxes, and date/time selectors. All of these are used in every app in cases of register or login flow, as well as the purchase form. The experience of creating such forms is very valuable and this chapter will help you to know how to create any form in your future apps. You'll learn the differences between iOS and Android and how to implement the appropriate abstractions for your app.

The following topics will be covered in this chapter:

- Collecting text input
- Selecting from a list of options
- Toggling between on and off
- Collecting date/time input

Technical requirements

You can find the code files for this chapter on GitHub at `https://github.com/PacktPublishing/React-and-React-Native-4th-Edition/tree/main/Chapter22`.

Collecting text input

It turns out that there's a lot to think about when it comes to implementing text inputs. For example, should it have placeholder text? Is this sensitive data that shouldn't be displayed on the screen? Should you process text as it's entered or when the user moves to another field?

The noticeable difference between mobile text input and traditional web text input is that the former has its own built-in virtual keyboard that you can configure and respond to. Let's build an example that renders several instances of the `<TextInput>` component:

```
import React, { useState } from "react";
import PropTypes from "prop-types";
import { Text, TextInput, View } from "react-native";
import styles from "./styles";

function Input(props) {
  return (
    <View style={styles.textInputContainer}>
      <Text style={styles.textInputLabel}>{props.label}
        </Text>
      <TextInput style={styles.textInput} {...props} />
    </View>
  );
}

Input.propTypes = {
  label: PropTypes.string,
};
```

We have implemented the `Input` component that we will reuse several times. Let's take a look at a few use cases of text inputs:

```
export default function CollectingTextInput() {
  const [changedText, setChangedText] = useState("");
  const [submittedText, setSubmittedText] = useState("");

  return (
    <View style={styles.container}>
      <Input label="Basic Text Input:" />
      <Input label="Password Input:" secureTextEntry />
      <Input label="Return Key:" returnKeyType="search" />
      <Input label="Placeholder Text:"
        placeholder="Search" />
      <Input
        label="Input Events:"
        onChangeText={(e) => {
          setChangedText(e);
        }}
        onSubmitEditing={(e) => {
          setSubmittedText(e.nativeEvent.text);
        }}
        onFocus={() => {
          setChangedText("");
          setSubmittedText("");
        }}
      />
      <Text>Changed: {changedText}</Text>
      <Text>Submitted: {submittedText}</Text>
    </View>
  );
}
```

I won't go into depth about what each of these `<TextInput>` components is doing – there are comments in the code that explain this. Let's see what these components look like on the screen:

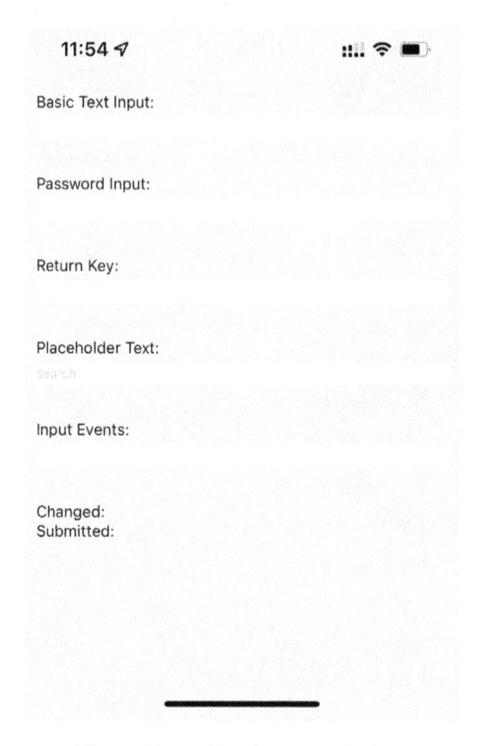

Figure 22.1 – Text input variations

The plain text input shows the text that's been entered. The **Password Input** field doesn't reveal any characters. **Placeholder Text** is displayed when the input is empty. The **Changed** text state is also displayed. You can't see the **Submitted** text state because I didn't press the **Submitted** button on the virtual keyboard before I took the screenshot.

Let's take a look at the virtual keyboard for the input element where you changed the **Return Key** text via the `returnKeyType` prop:

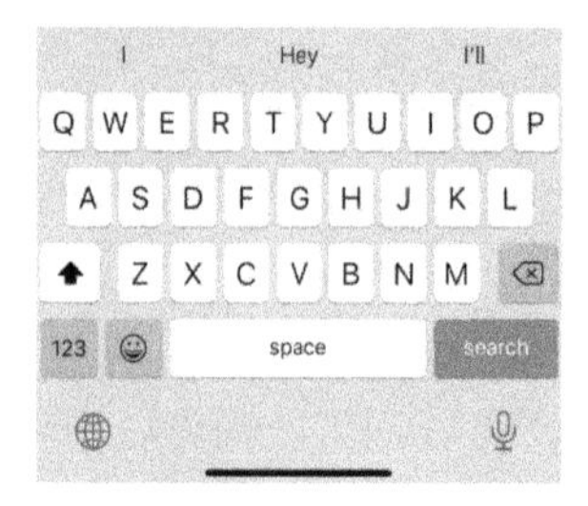

Figure 22.2 – Keyboard with changed Return key text

When the keyboard *Return* key reflects what's going to happen when the user presses it, the user feels more in tune with the application.

One more common use case is changing the keyboard type. By providing the `keyboardType` prop to the `TextInput` component, you will see different variations of keyboards. This is convenient when you need to enter a pin code or email address. Here is an example of a numeric keyboard:

Figure 22.3 – Numeric keyboard type

Now that you're familiar with collecting text input, it's time to learn how to select a value from a list of options.

Selecting from a list of options

In web applications, you typically use the `<select>` element to let the user choose from a list of options. React Native comes with a `<Picker>` component, which works on both iOS and Android, but in terms of reducing the React Native app size, the Meta team decided to delete it in future releases and extract Picker to its own package. To use that package, firstly, we install it in a clean project by running this command:

```
expo install @react-native-picker/picker
```

There is some trickery involved with styling this component based on which platform the user is on, so let's hide all of this inside a generic `Select` component. Here's the `Select.ios.js` module:

```
import React from "react";
import { View, Text } from "react-native";
import { Picker } from "@react-native-picker/picker";
import styles from "./styles";

export default function Select(props) {
  return (
    <View style={styles.pickerHeight}>
```

```
        <View style={styles.pickerContainer}>
          <Text style={styles.pickerLabel}>{props.label}
            </Text>
          <Picker style={styles.picker} {...props}>
            {props.items.map((i) => (
              <Picker.Item key={i.label} {...i} />
            ))}
          </Picker>
        </View>
      </View>
    );
}
```

That's a lot of overhead for a simple `Select` component. Well, it turns out that it's actually quite hard to style the React Native `<Picker>` component. Here's the `Select.android.js` module:

```
import React from "react";
import { View, Text } from "react-native";
import { Picker } from "@react-native-picker/picker";
import styles from "./styles";

export default function Select(props) {
  return (
    <View>
      <Text style={styles.pickerLabel}>{props.label}</Text>
      <Picker {...props}>
        {props.items.map((i) => (
          <Picker.Item key={i.label} {...i} />
        ))}
      </Picker>
    </View>
  );
}
```

This is what the styles look like:

```
import { StyleSheet } from "react-native";

export default StyleSheet.create({
  container: {
    flex: 1,
    flexDirection: "column",
    backgroundColor: "ghostwhite",
    justifyContent: "center",
  },

  pickersBlock: {
    flex: 2,
    flexDirection: "row",
    justifyContent: "space-around",
    alignItems: "center",
  },
```

As usual with the `container` and `pickersBlock` styles, we define the base layout of the screen. Next, let's take a look at the styles of the `<Select>` component:

```
  pickerHeight: {
    height: 250,
  },

  pickerContainer: {
    flex: 1,
    flexDirection: "column",
    alignItems: "center",
    backgroundColor: "white",
    padding: 6,
    height: 240,
  },

  pickerLabel: {
    fontSize: 14,
```

```
    fontWeight: "bold",
  },

  picker: {
    width: 150,
    backgroundColor: "white",
  },

  selection: {
    flex: 1,
    textAlign: "center",
  },
});
```

Now, you can render your `<Select>` component. Here is what the `App.js` file looks like:

```
const sizes = [
  { label: "", value: null },
  { label: "S", value: "S" },
  { label: "M", value: "M" },
  { label: "L", value: "L" },
  { label: "XL", value: "XL" },
];
const garments = [
  { label: "", value: null, sizes: ["S", "M", "L", "XL"] },
  { label: "Socks", value: 1, sizes: ["S", "L"] },
  { label: "Shirt", value: 2, sizes: ["M", "XL"] },
  { label: "Pants", value: 3, sizes: ["S", "L"] },
  { label: "Hat", value: 4, sizes: ["M", "XL"] },
];
```

Here, we defined the default values for our `<Select/>` component. Let's take a look at the final `SelectingOptions` component:

```
export default function SelectingOptions() {
  const [availableGarments, setAvailableGarments] =
    useState([]);
  const [selectedSize, setSelectedSize] = useState(null);
```

```
const [selectedGarment, setSelectedGarment] =
  useState(null);
const [selection, setSelection] = useState("");
```

With these hooks, we've implemented states of selectors. Next, we will use and pass them into components:

```
return (
  <View style={styles.container}>
    <View style={styles.pickersBlock}>
      <Select
        label="Size"
        items={sizes}
        selectedValue={selectedSize}
        onValueChange={(size) => {
          setSelectedSize(size);
          setSelectedGarment(null);
          setAvailableGarments(
            garments.filter((i) =>
              i.sizes.includes(size))
          );
        }}
      />
      <Select
        label="Garment"
        items={availableGarments}
        selectedValue={selectedGarment}
        onValueChange={(garment) => {
          setSelectedGarment(garment);
          setSelection(
            '${selectedSize} ${
              garments.find((i) => i.value ===
                garment).label
            }'
          );
        }}
      />
```

```
        </View>
        <Text style={styles.selection}>{selection}</Text>
      </View>
    );
}
```

The basic idea of this example is that the selected option in the first selector changes the available options in the second selector. When the second selector changes, the label shows `selectedSize` and `selectedGarment` as a string. Here's how the screen looks:

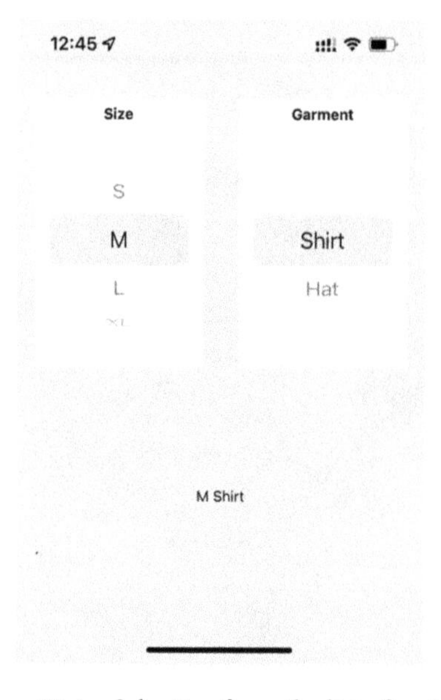

Figure 22.4 – Selecting from the list of options

The size selector is shown on the left-hand side of the screen. When the size value changes, the available values in the garment selector on the right-hand side of the screen change to reflect size availability. The current selection is displayed as a string after the two selectors. In the following section, you'll learn about the buttons that toggle between on and off states.

Toggling between on and off

Another common element you'll see in web forms is checkboxes. React Native has a `Switch` component that works on both iOS and Android. Thankfully, this component is a little easier to style than the `Picker` component. Let's look at a simple abstraction you can implement to provide labels for your switches:

```
import React from "react";
import { View, Text, Switch } from "react-native";
import styles from "./styles";

export default function CustomSwitch(props) {
  return (
    <View style={styles.customSwitch}>
      <Text>{props.label}</Text>
      <Switch {...props} />
    </View>
  );
}
```

Now, let's learn how we can use a couple of switches to control application state:

```
import React, { useState } from "react";
import { View } from "react-native";
import styles from "./styles";
import Switch from "./Switch";

export default function TogglingOnAndOff() {
  const [first, setFirst] = useState(false);
  const [second, setSecond] = useState(false);

  return (
    <View style={styles.container}>
      <Switch
        label="Disable Next Switch"
        value={first}
        disabled={second}
        onValueChange={setFirst}
```

```
      />
      <Switch
        label="Disable Previous Switch"
        value={second}
        disabled={first}
        onValueChange={setSecond}
      />
    </View>
  );
}
```

These two switches toggle the `disabled` property of one another. When the first switch is toggled, the `setFirst()` function is called, which will update the value of the `first` state. Depending on the current value of `first`, it will either be set to `true` or `false`. The second switch works the same way, except it uses `setSecond()` and the `second` state value.

Turning on one switch will disable the other because we've set the `disabled` property value for each switch to the state of the other switch. For example, the second switch has `disabled={first}`, which means that it is disabled whenever the first switch is turned on. Here's what the screen looks like on iOS:

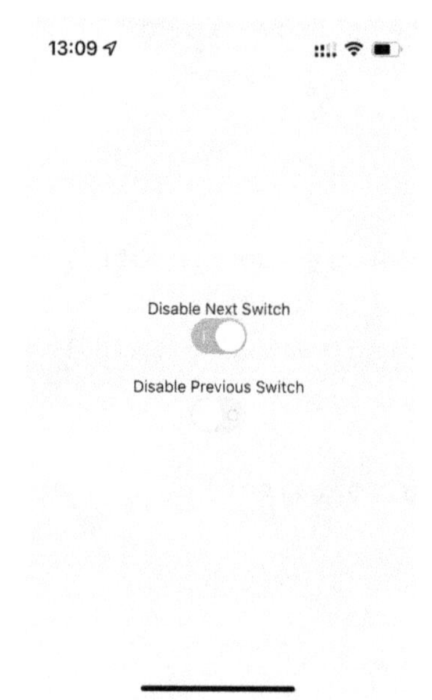

Figure 22.5 – Switch toggles on iOS

Here's what the same screen looks like on Android:

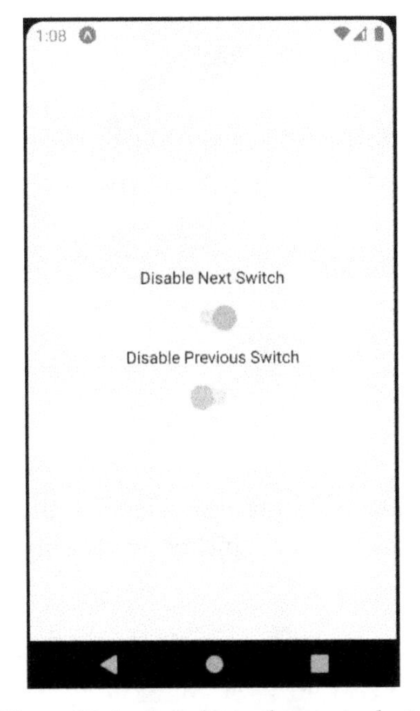

Figure 22.6 – Switch toggles on Android

As you can see, our `CustomSwitch` component enables the same functionality on Android and iOS while using one component for both platforms. In the following section, you'll learn how to collect date/time input.

Collecting date/time input

In this final section of this chapter, you'll learn how to implement date/time pickers. React Native docs suggest using `@react-native-community/datetimepicker` independent date/time picker components for iOS and Android, which means that it is up to you to handle the cross-platform differences between the components. To install `datetimepicker`, run the following command in the project:

```
expo install @react-native-community/datetimepicker
```

So, let's start with a date picker component for iOS:

```
import React from "react";
import { Text, View } from "react-native";
import DateTimePicker from "@react-native-
  community/datetimepicker";
```

```
import styles from "./styles";

export default function DatePicker(props) {
  return (
    <View style={styles.datePickerContainer}>
      <Text style={styles.datePickerLabel}>{props.label}
        </Text>
      <DateTimePicker mode="date" display="spinner"
        {...props} />
    </View>
  );
}
```

There's not a lot to this component; it simply adds a label to the `DateTimePicker` component. The Android version of the date picker needs a little more work. Let's take a look at the implementation:

```
import React from "react";
import { Text, View } from "react-native";
import DateTimePicker from "@react-native-
  community/datetimepicker";
import styles from "./styles";

function pickDate(options, onDateChange) {
  DateTimePicker.open(options).then((date) =>
    onDateChange(new Date(date.year, date.month, date.day))
  );
}

export default function DatePicker({ label, date,
  onDateChange }) {
  return (
    <View style={styles.datePickerContainer}>
      <Text style={styles.datePickerLabel}>{label}</Text>
      <Text onPress={() => pickDate({ date },
        onDateChange)}>
```

```
        {date.toLocaleDateString()}
      </Text>
    </View>
  );
}
```

The key difference between the two date pickers is that the Android version doesn't use a React Native component, such as `DateTimePicker` in iOS. Instead, we have to use the imperative `DateTimePicker.open()` API. This is triggered when the user presses the date text that our component renders and opens a date picker dialog. The good news is that this component of ours hides this API behind a declarative component.

> **Important Note**
> I've also implemented a time picker component that follows this exact pattern. So, rather than listing that code here, I suggest that you download the code for this book from `https://github.com/PacktPublishing/React-and-React-Native-4th-Edition-` so that you can see the subtle differences and run the example.

Now, let's learn how to use our date and time picker components:

```
import React, { useState } from "react";
import { View } from "react-native";
import DatePicker from "./DatePicker";
import TimePicker from "./TimePicker";
import styles from "./styles";

export default function CollectingDateTimeInput() {
  const [date, setDate] = useState(new Date());
  const [time, setTime] = useState(new Date());

  return (
    <View style={styles.container}>
      <DatePicker
        label="Pick a date, any date:"
        value={date}
        onChange={setDate}
      />
```

```
      <TimePicker
        label="Pick a time, any time:"
        value={time}
        onChange={setTime}
      />
    </View>
  );
}
```

Awesome! Now, we have `DatePicker` and `TimePicker` components that can help us to select dates and times in our app. Also, they both work on iOS and Android. Let's see how the pickers look on iOS:

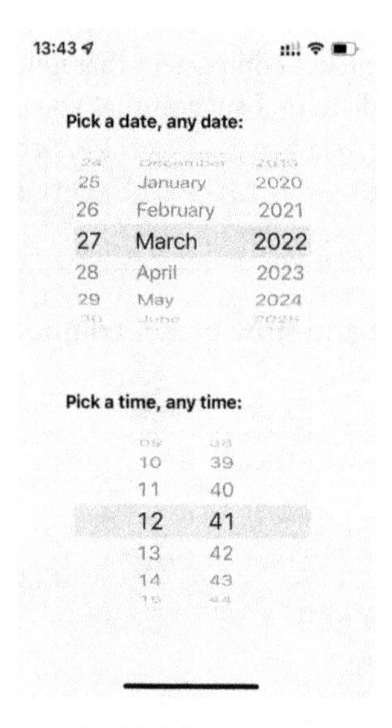

Figure 22.7 – iOS date and time pickers

As you can see, the iOS date and time pickers use the `Picker` component that you learned about earlier in this chapter. The Android picker looks a lot different – let's look at it now:

Figure 22.8 – Android date picker

The Android version follows a completely different approach from the iOS date/time picker, yet we can use the same `DatePicker` component that we've created on both platforms. This brings us to the end of the chapter.

Summary

In this chapter, we learned about the various React Native components that resemble the form elements from the web that we're used to. We started off by learning about text input and how each text input has its own virtual keyboard to take into consideration. Next, we learned about `Picker` components, which allow the user to select an item from a list of options. Then, we learned about the `Switch` component, which is kind of like a checkbox. With these components, you will be able to build a form of any complexity.

In the final section, we learned how to implement generic date/time pickers that work on both iOS and Android. In the next chapter, we'll learn about modal dialogs in React Native.

Further reading

Visit the following links for more information:

- Handling text input: `https://reactnative.dev/docs/handling-text-input`
- `Switch`: `https://reactnative.dev/docs/switch`
- `Picker`: `https://docs.expo.dev/versions/latest/sdk/picker/`
- `DateTimePicker`: `https://docs.expo.dev/versions/latest/sdk/date-time-picker/`

23 Displaying Modal Screens

The goal of this chapter is to show you how to present information to the user in ways that don't disrupt the current page. Pages use a `View` component and render it directly on the screen. There are times, however, when there's important information that the user needs to see but you don't necessarily want to kick them off the current page.

You'll start by learning how to display important information. By knowing what information is important and when to use it, you'll learn how to get user acknowledgment – both for error and success scenarios. Then, you'll implement passive notifications that show the user that something has happened. Finally, you'll implement modal views that show the user that something is happening in the background.

The following topics will be covered in this chapter:

- Important information
- Getting user confirmation
- Passive notifications
- Activity modals

Technical requirements

You can find the code files for this chapter on GitHub at `https://github.com/PacktPublishing/React-and-React-Native-4th-Edition/tree/main/Chapter23`.

Important information

Before you dive into implementing alerts, notifications, and confirmations, let's take a few minutes and think about what each of these items means. I think this is important because if you end up passively notifying the user about an error, it can easily get missed. Here are my definitions of the types of information that you'll want to display:

- **Alert**: Something important just happened, and you need to ensure that the user sees what's going on. Possibly, the user needs to acknowledge the alert.
- **Confirmation**: This is part of an alert. For example, if the user has just performed an action and then wants to make sure that it was successful before carrying on, they would have to confirm that they've seen the information in order to close the modal. A confirmation can also exist within an alert, warning the user about an action that they're about to perform.
- **Notification**: Something happened but it's not important enough to completely block what the user is doing. These typically go away on their own.

The trick is to try to use notifications where the information is good to know but not critical. Use confirmations only when the workflow of the feature cannot continue without the user acknowledging what's going on. In the following sections, you'll see examples of alerts and notifications that are used for different purposes.

Getting user confirmation

In this section, you'll learn how to show modal views in order to get confirmation from the user. First, you'll learn how to implement a successful scenario, where an action generates a successful outcome that you want the user to be aware of. Then, you'll learn how to implement an error scenario where something went wrong and you don't want the user to move forward without acknowledging the issue.

Displaying a success confirmation

Let's start by implementing a modal view that's displayed as a result of the user successfully performing an action. Here's the `Modal` component, which is used to show the user a confirmation modal:

```
iexport default function ConfirmationModal(props) {
  return (
    <Modal {...props}>
      <View style={styles.modalContainer}>
        <View style={styles.modalInner}>
          <Text style={styles.modalText}>Dude,
            srsly?</Text>
          <Text style={styles.modalButton}
            onPress={props.onPressConfirm}>
            Yep
          </Text>
          <Text style={styles.modalButton}
            onPress={props.onPressCancel}>
            Nope
          </Text>
        </View>
      </View>
    </Modal>
  );
}

ConfirmationModal.defaultProps = {
  transparent: true,
  onRequestClose: () => {},
};
```

The properties that are passed to `ConfirmationModal` are forwarded to the React Native `Modal` component. You'll see why in a moment. First, let's see what this confirmation modal looks like:

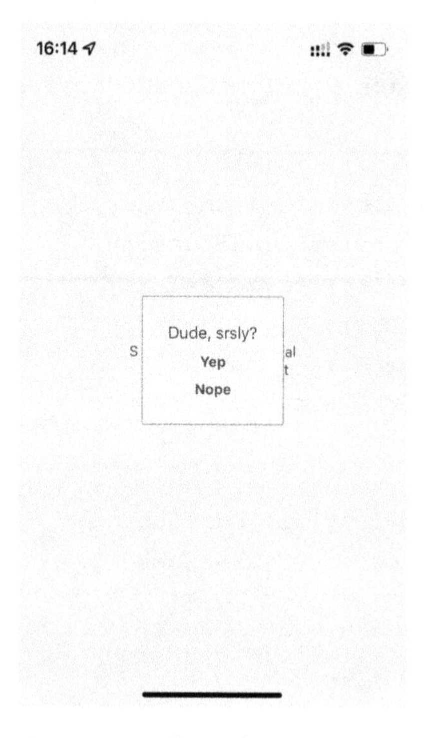

Figure 23.1 – The confirmation modal

The modal that's displayed once the user completes an action uses our own styling and confirmation message. It also has two actions, but it may only need one, depending on whether this confirmation is pre-action or post-action. Here are `styles` that are being used for this modal:

```
modalContainer: {
    flex: 1,
    justifyContent: "center",
    alignItems: "center",
  },

  modalInner: {
    backgroundColor: "azure",
    padding: 20,
```

```
    borderWidth: 1,
    borderColor: "lightsteelblue",
    borderRadius: 2,
    alignItems: "center",
  },

  modalText: {
    fontSize: 16,
    margin: 5,
    color: "slategrey",
  },

  modalButton: {
    fontWeight: "bold",
    margin: 5,
    color: "slategrey",
  },
```

With the React Native `Modal` component, it's pretty much up to you how you want your confirmation modal view to look. Think of them as regular views, with the only difference being that they're rendered on top of other views.

A lot of the time, you might not care to style your own modal views. For example, in web browsers, you can simply call the `alert()` function, which shows text in a window that's styled by the browser. React Native has something similar – `Alert.alert()`. The tricky part here is that this is an imperative API, and you don't necessarily want to expose it directly to your application.

Instead, let's implement a `ConfirmationAlert` component that hides the details of this particular React Native API so that your app can just treat this like any other component:

```
import React from "react";
import { Alert } from "react-native";

export default function ConfirmationAlert(props) {
  React.useEffect(() => {
    if (props.visible) {
      Alert.alert(props.title, props.message,
        props.buttons);
```

```
    }
  }, [props.visible]);

  return null;
}

ConfirmationAlert.defaultProps = {
  title: "",
  message: "",
  buttons: [],
};
```

This component doesn't need to render anything, since it deals exclusively with imperative React Native calls. However, it feels like something is being rendered to the person that's using `ConfirmationAlert`.

Here's what the alert looks like on iOS:

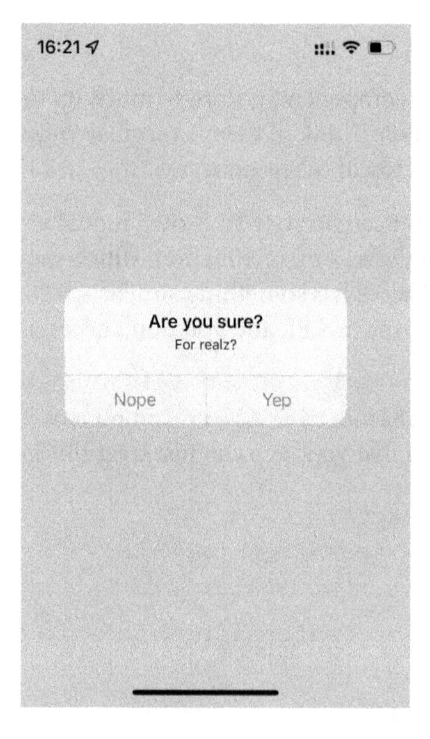

Figure 23.2 – A confirmation alert on iOS

In terms of functionality, there's nothing really different here. There is a title and text beneath it, but that's something that can easily be added to a modal view if you wanted. The real difference is that this modal looks like an iOS modal instead of something that's styled by the app. Let's see how this alert appears on Android:

Figure 23.3 – A confirmation alert on Android

This modal looks like an Android modal, and you didn't have to style it. I think using alerts over modals is a better choice most of the time. It makes sense to have something styled to look like it's part of iOS or Android. However, there are times when you need more control over how the modal looks, such as when displaying error confirmations.

The approach to rendering modals is different from the approach to rendering alerts. However, they're both still declarative components that change based on the changing property values.

Error confirmation

All of the principles you learned about in the *Displaying a success confirmation* section are applicable when you need the user to acknowledge an error. If you need more control of the display, use a modal. For example, you might want the modal to be red and scary-looking, like this:

Figure 23.4 – The error confirmation modal

Here are `styles` that were used to create this look. Maybe you want something a bit more subtle, but the point is that you can make this look however you want:

```
modalInner: {
  backgroundColor: "azure",
  padding: 20,
  borderWidth: 1,
  borderColor: "lightsteelblue",
  borderRadius: 2,
  alignItems: "center",
},
```

In the `modalInner` style property, we've defined screen styles. Next, we'll define modal styles:

```
  modalInnerError: {
    backgroundColor: "lightcoral",
    borderColor: "darkred",
  },
  modalText: {
    fontSize: 16,
    margin: 5,
    color: "slategrey",
  },
  modalTextError: {
    fontSize: 18,
    color: "darkred",
  },
  modalButton: {
    fontWeight: "bold",
    margin: 5,
    color: "slategrey",
  },
  modalButtonError: {
    color: "black",
  },
```

The same modal styles that you used for the success confirmations are still here. That's because the error confirmation modal needs many of the same style properties. Here's how you apply both to the `Modal` component:

```
import React from "react";
import { View, Text, Modal } from "react-native";
import styles from "./styles";

const innerViewStyle = [styles.modalInner,
  styles.modalInnerError];
const textStyle = [styles.modalText,
  styles.modalTextError];
const buttonStyle = [styles.modalButton,
```

```
    styles.modalButtonError];
export default function ErrorModal(props) {
  return (
    <Modal {...props}>
      <View style={styles.modalContainer}>
        <View style={innerViewStyle}>
          <Text style={textStyle}>Epic fail!</Text>
          <Text style={buttonStyle}
            onPress={props.onPressConfirm}>
            Fix it
          </Text>
          <Text style={buttonStyle}
            onPress={props.onPressCancel}>
            Ignore it
          </Text>
        </View>
      </View>
    </Modal>
  );
}
```

The styles are combined as arrays before they're passed to the `style` component property. The styles error always comes last, since conflicting style properties, such as `backgroundColor`, will be overridden by whatever comes later in the array.

In addition to styles in error confirmations, you can include whatever advanced controls you want. It really depends on how your application lets users cope with errors – for example, maybe there are several courses of action that can be taken.

However, the more common case is that something went wrong, and there's nothing you can do about it besides making sure that the user is aware of the situation. In these cases, you can probably get away with just displaying an alert:

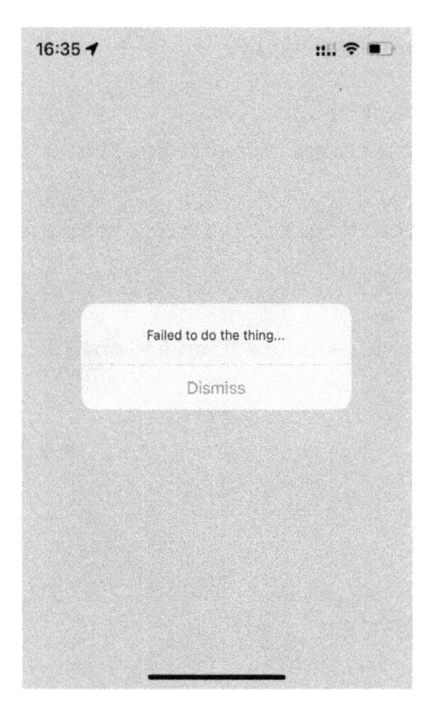

Figure 23.5 – An error alert

Now that you're able to display error notifications that require user engagement, it's time to learn about less aggressive notifications that don't disrupt what the user is currently doing.

Passive notifications

The notifications you've examined so far in this chapter all have required input from the user. This is by design because it's important information that you're forcing the user to look at. However, you don't want to overdo this. For notifications that are important but not life-altering if ignored, you can use passive notifications. These are displayed in a less obtrusive way than modals and don't require any user action to dismiss them.

In this section, you'll create a `Notification` component that uses the Toast API for Android and creates a custom modal for iOS. It's called the Toast API because the information that's displayed looks like a piece of toast popping up. `Toasts` is a common component in Android to show some basic information that does not require user response.

Here's what the Android component looks like:

```
import React from "react";
import { ToastAndroid } from "react-native";

export default function Notification({ message, duration }) {
  React.useEffect(() => {
    if (message) {
      ToastAndroid.show(message, duration);
    }
  }, [message]);

  return null;
}

Notification.defaultProps = {
  duration: ToastAndroid.LONG,
};
```

Once again, you're dealing with an imperative React Native API that you don't want to expose to the rest of your app. Instead, this component hides the imperative `ToastAndroid.show()` function behind a declarative React component. No matter what, this component returns `null` because it doesn't actually render anything. Here's what the `ToastAndroid` notification looks like:

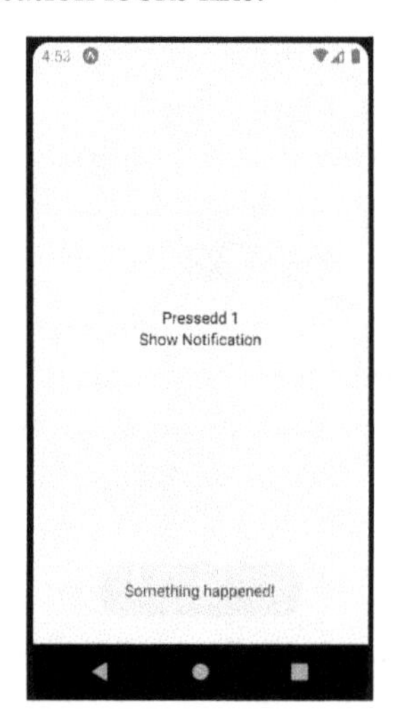

Figure 23.6 – An Android notification

A notification stating **Something happened!** is displayed at the bottom of the screen and is removed after a short delay. The key is that the notification is unobtrusive.

The iOS notification component is a little more involved because it needs state and life cycle events to make a modal view behave like a transient notification. Here's what the code for it looks like:

```
export default function Notification(props) {
  const [message, setMessage] = useState(props.message);
  useEffect(() => {
    setMessage(props.message);

    const timer = setTimeout(() => {
      setMessage(null);
    }, props.duration);

    return () => {
      clearTimeout(timer);
    };
  }, [props.message]);

  const modalProps = {
    animationType: "fade",
    transparent: true,
    visible: Boolean(message),
  };
```

The logic in the preceding code block describes how to render a message that disappears after a duration period. In the following code block, you can see how we applied this logic to the layout:

```
  return (
    <Modal {...modalProps}>
      <View style={styles.notificationContainer}>
        <View style={styles.notificationInner}>
          <Text>{message}</Text>
        </View>
      </View>
```

```
    </Modal>
  );
}
```

You have to style the modal to display the notification text, as well as the state that's used to hide the notification after a delay. Here's what the end result looks like for iOS:

Figure 23.7 – An iOS notification

The same principle for the `ToastAndroid` API applies here. You might have noticed that there's another button in addition to the `Show Notification` button. This is a simple counter that re-renders the view. There's actually a reason for demonstrating this seemingly obtuse feature, as you'll see momentarily. Here's the code for the main application view:

```
export default function PassiveNotifications() {
  const [count, setCount] = useState(0);
  const [message, setMessage] = useState(null);

  return (
    <View style={styles.container}>
      <Notification message={message} />
```

```
        <Text
          onPress={() => {
            setCount(count + 1);
            setMessage(null);
          }}
        >
          Pressed {count}
        </Text>
        <Text
          onPress={() => {
            setMessage("Something happened!");
          }}
        >
          Show Notification
        </Text>
      </View>
    );
}
```

The whole point of the press counter is to demonstrate that even though the `Notification` component is declarative and accepts new property values when the state changes, you still have to set the message state to `null` when changing other state values. The reason for this is that if you re-render the component and the message state still has a string in it, it will display the same notification over and over.

In the next section, you'll learn about activity modals, which show the user that something is happening.

Activity modals

In this final section of this chapter, you'll implement a modal that shows a progress indicator. The idea is to display the modal and then hide it when the promise resolves. Here's the code for the generic `Activity` component, which shows a modal with `ActivityIndicator`:

```
import React from "react";
import { View, Modal, ActivityIndicator } from
  "react-native";
import styles from "./styles";
```

```
export default function Activity(props) {
  return (
    <Modal visible={props.visible} transparent>
      <View style={styles.modalContainer}>
        <ActivityIndicator size={props.size} />
      </View>
    </Modal>
  );
}

Activity.defaultProps = {
  visible: false,
  size: "large",
};
```

You might be tempted to pass the promise to the component so that it automatically hides when the promise resolves. I don't think this is a good idea because then you would have to introduce the state into this component. Furthermore, it would depend on a promise in order to function. With the way you've implemented this component, you can show or hide the modal based on the `visible` property alone. Here's what the activity modal looks like on iOS:

Figure 23.8 – An activity modal

There's a semi-transparent background on the modal that's placed over the main view with the **Fetch Stuff...** link. By clicking on this link, we will be shown the activity loader. Here's how this effect is created in `styles.js`:

```
modalContainer: {
    flex: 1,
    justifyContent: "center",
    alignItems: "center",
    backgroundColor: "rgba(0, 0, 0, 0.2)",
  },
```

Instead of setting the actual `Modal` component to transparent, you can set the transparency in `backgroundColor`, which gives the look of an overlay. Now, let's take a look at the code that controls this component:

```
export default function App() {
  const [fetching, setFetching] = useState(false);
  const [promise, setPromise] =
    useState(Promise.resolve());
  function onPress() {
    setPromise(
      new Promise((resolve) => setTimeout(resolve,
        3000)).then(() => {
        setFetching(false);
      })
    );
    setFetching(true);
  }
  return (
    <View style={styles.container}>
      <Activity visible={fetching} />
      <Text onPress={onPress}>Fetch Stuff...</Text>
    </View>
  );
}
```

When the fetch link is pressed, a new promise is created that simulates asynchronous network activity. Then, when the promise resolves, you can change the `fetching` state back to `false` so that the activity dialog is hidden.

Summary

In this chapter, we learned about the need to show important information to mobile users. This sometimes involves explicit feedback from the user, even if that just means acknowledging the message. In other cases, passive notifications work better, since they're less obtrusive than confirmation modals.

There are two tools that we can use to display messages to users – modals and alerts. Modals are more flexible because they're just like regular views. Alerts are good for displaying plain text, and they take care of styling concerns for us. On Android, we have the `ToastAndroid` interface as well. We saw that it's also possible to do this on iOS, but it just requires more work.

In the next chapter, we'll dig deeper into the gesture response system inside React Native, which makes for a better mobile experience than browsers can provide.

Further reading

Check out the following links for more information:

- `Modal`: `https://reactnative.dev/docs/modal`
- `Alert`: `https://reactnative.dev/docs/alert`
- `ToastAndroid`: `https://reactnative.dev/docs/toastandroid`

24
Responding to User Gestures

All of the examples that you've implemented so far in this book have relied on user gestures. In traditional web applications, you mostly deal with mouse events. However, touchscreens rely on the user manipulating elements with their fingers, which is fundamentally different from the mouse.

In this chapter, first, you'll learn about scrolling. This is probably the most common gesture, besides touch. Then, you'll learn about giving the user the appropriate level of feedback when they interact with your components. Finally, you'll implement components that can be swiped.

The goal of this chapter is to show you how the gesture response system inside React Native works and some of the ways in which this system is exposed via components.

In this chapter, we'll cover the following topics:

- Scrolling with your fingers
- Giving touch feedback
- Using Swipeable and cancellable components

Technical requirements

You can find the code files for this chapter on Github at `https://github.com/PacktPublishing/React-and-React-Native-4th-Edition/tree/main/Chapter24`.

Scrolling with your fingers

Scrolling in web applications is done by using the mouse pointer to drag the scrollbar back and forth or up and down, or by spinning the mouse wheel. This doesn't work on mobile devices because there's no mouse. Everything is controlled by gestures on the screen. For example, if you want to scroll down, you use your thumb or index finger to pull the content up by physically moving your finger over the screen.

Scrolling like this is difficult to implement, but it gets more complicated. When you scroll on a mobile screen, the velocity of the dragging motion is taken into consideration. You drag the screen fast, then let go, and the screen continues to scroll based on how fast you moved your finger. You can also touch the screen while this is happening to stop it from scrolling.

Thankfully, you don't have to handle most of this stuff. The `ScrollView` component handles much of the scrolling complexity for you. In fact, you've already used the `ScrollView` component back in *Chapter 19, Rendering Item Lists*. The `ListView` component has `ScrollView` baked into it.

> **Important Note**
>
> You can hack the low-level parts of user interactions by implementing gesture life cycle methods. You'll probably never need to do this, but if you're interested, you can read about it at `https://reactnative.dev/docs/gesture-responder-system`.

You can use `ScrollView` outside of `ListView`. For example, if you're just rendering arbitrary content such as text and other widgets – not a list, in other words – you can just wrap it in `<ScrollView>`. Here's an example:

```
import React from "react";
import {
  Text,
  ScrollView,
```

```
  ActivityIndicator,
  Switch,
  View,
} from "react-native";

import styles from "./styles";

export default function App() {
  return (
    <View style={styles.container}>
      <ScrollView style={styles.scroll}>
        {new Array(20).fill(null).map((v, i) => (
          <View key={i}>
            <Text style={[styles.scrollItem,
              styles.text]}>Some text</Text>
            <ActivityIndicator style={styles.scrollItem}
              size="large" />
            <Switch style={styles.scrollItem} />
          </View>
        ))}
      </ScrollView>
    </View>
  );
}
```

The `ScrollView` component isn't of much use on its own – it's there to wrap other components. It needs height in order to function correctly. Here's what the scroll style looks like:

```
scroll: {
    height: 1,
    alignSelf: "stretch",
  },
```

The `height` property is set to `1`, but the `stretch` value of `alignSelf` allows the items to display properly. Here's what the end result looks like:

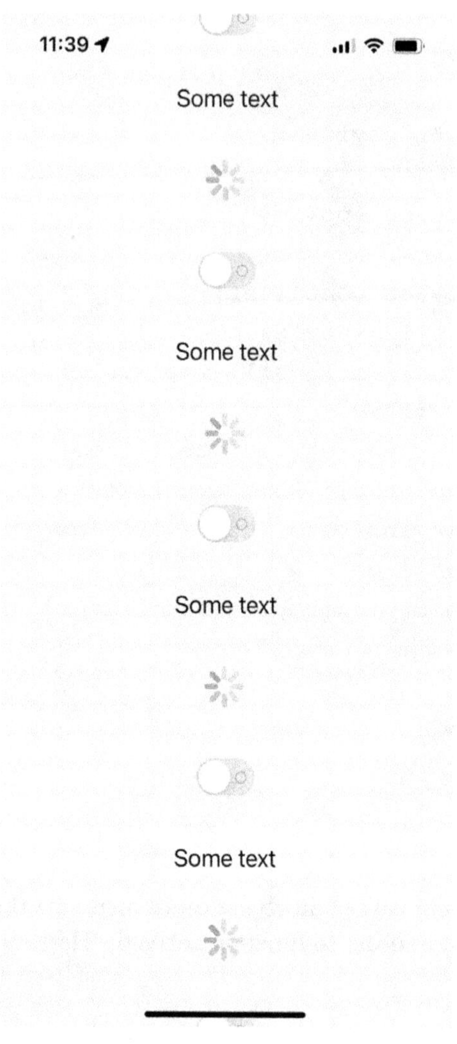

Figure 24.1 – ScrollView

There's a vertical scrollbar on the right-hand side of the screen as you drag the content down. If you run this example, you can play around with making various gestures, such as making content scroll on its own and then making it stop.

When the user scrolls through content on the screen, they receive visual feedback. Users should also receive visual feedback when they touch certain elements on the screen.

Giving touch feedback

The React Native examples you've worked with so far in this book have used plain text to act as buttons or links. In web applications, to make text look like something that can be clicked, you just wrap it with the appropriate link. There's no such thing as mobile links, so you can style your text to look like a button.

> **Important Note**
>
> The problem with trying to style text as links on mobile devices is that they're too hard to press. Buttons provide a bigger target for fingers, and they're easier to apply touch feedback on.

Let's style some text as a button. This is a great first step as it makes the text look touchable. But you also want to give visual feedback to the user when they start interacting with the button. React Native provides several components to help with this:

- `TouchableOpacity`
- `TouchableHighlight`
- `Pressable API`

But before diving into the code, let's take a look at what these components look like visually when users interact with them, starting with `TouchableOpacity`:

Figure 24.2 – TouchableOpacity

There are three buttons being rendered here. The top one, labeled **Opacity**, is currently being pressed by the user. The opacity of the button is dimmed when pressed, which provides important visual feedback for the user. Let's see what the **Highlight** button looks like when pressed:

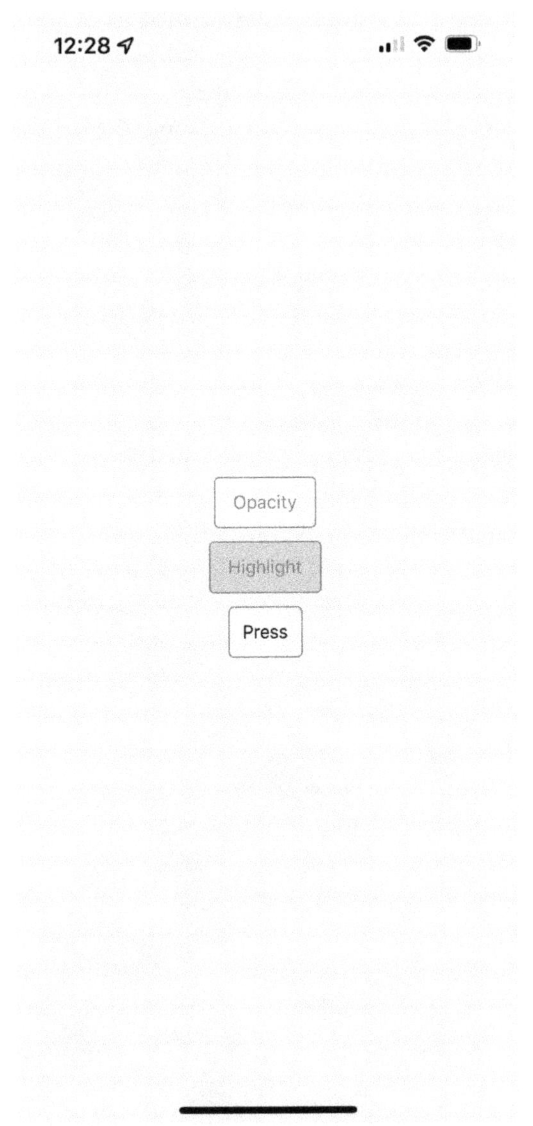

Figure 24.3 – TouchableHighlight

Instead of changing the opacity when pressed, the `TouchableHighlight` component adds a highlight layer over the button. In this case, it's highlighted using a more transparent version of the slate gray that's being used in the font and border colors.

The last example of a button is provided by the `Pressable` component. The Pressable API has been introduced as a core component wrapper and allows different stages of press interaction on any of its defined children. With such components, we can handle `onPressIn`, `onPressOut`, and `onLongPress` callbacks and implement any touchable feedback that we want. Let's take a look at how `PressableButton` looks when we click on it:

Figure 24.4 – Pressable button

If we continue to keep our finger on this button, we get an `onLongPress` event and the button will update:

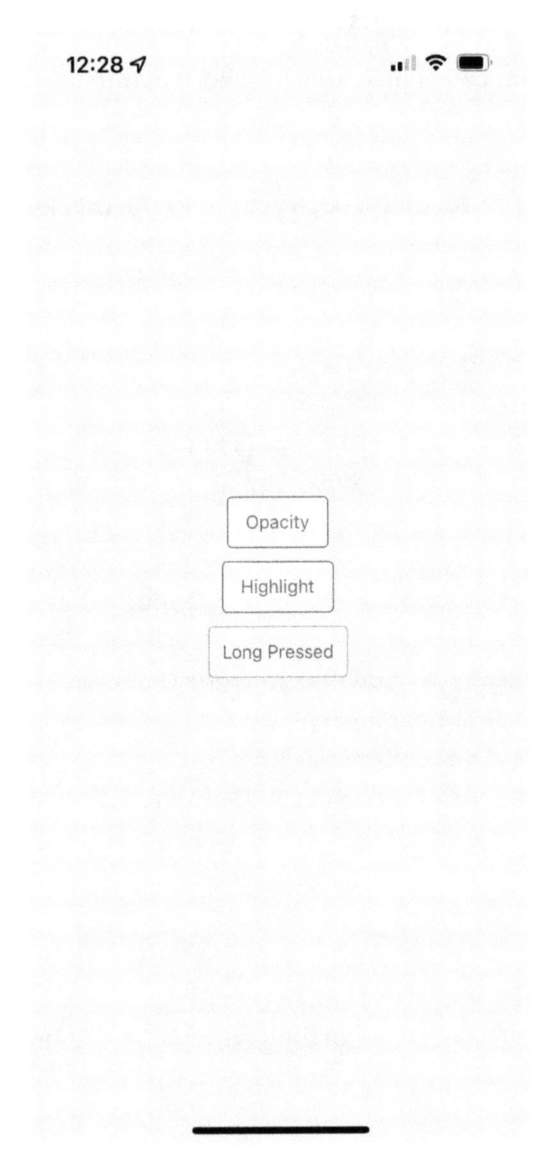

Figure 24.5 – Long Pressed button

It doesn't really matter which approach you use. The important thing is that you provide the appropriate touch feedback for your users as they interact with your buttons. In fact, you might want to use all the approaches in the same app, but for different things:

1. Let's create a `Button` component, which makes it easy to use the first two approaches:

```
import React from "react";
import { Text, TouchableOpacity, TouchableHighlight }
  from "react-native";
import styles from "./styles";

const touchables = new Map([
  ["opacity", TouchableOpacity],
  ["highlight", TouchableHighlight],
  [undefined, TouchableOpacity],
]);

export default function Button({ label, onPress,
  touchable }) {
  const Touchable = touchables.get(touchable);
  const touchableProps = {
    style: styles.button,
    underlayColor: "rgba(112,128,144,0.3)",
    onPress,
  };

  return (
    <Touchable {...touchableProps}>
      <Text style={styles.buttonText}> {label} </Text>
    </Touchable>
  );
}
```

The `touchables` map is used to determine which React Native touchable component wraps the text, based on the `touchable` property value. Here are the styles that were used to create this button:

```
button: {
    padding: 10,
    margin: 5,
    backgroundColor: "azure",
    borderWidth: 1,
    borderRadius: 4,
    borderColor: "slategrey",
  },

  buttonText: {
    color: "slategrey",
  },
```

2. Let's take a look at the button based on the Pressable API:

```
import React, { useState } from "react";
import { Pressable, Text } from "react-native";
import styles from "./styles";

export default PressableButton = () => {
  const [text, setText] = useState(0);

  return (
    <Pressable
      onPressIn={() => setText("Pressed")}
      onPressOut={() => setText("Press")}
      onLongPress={() => {
        setText("Long Pressed");
      }}
      delayLongPress={500}
      style={({ pressed }) => [
        {
          opacity: pressed ? 0.5 : 1,
        },
```

```
        styles.button,
      ]}
    >
      <Text style={styles.text}>{text}</Text>
    </Pressable>
  );
};
```

3. Here's how you can put those buttons into the main app module:

```
import React from "react";
import { View } from "react-native";
import styles from "./styles";
import Button from "./Button";
import PressableButton from "./PressableButton";

export default function App() {
  return (
    <View style={styles.container}>
      <Button onPress={() => {}} label="Opacity" />
      <Button onPress={() => {}} label="Highlight"
        touchable="highlight" />
      <PressableButton />
    </View>
  );
}
```

Note that the `onPress` callbacks don't actually do anything – we're passing them because they're a required property.

In the following section, you'll learn about providing feedback when the user swipes elements across the screen.

Using Swipeable and cancellable components

Part of what makes native mobile applications easier to use than mobile web applications is that they feel more intuitive. Using gestures, you can quickly get a handle on how things work. For example, swiping an element across the screen with your finger is a common gesture, but the gesture has to be discoverable.

Let's say that you're using an app, and you're not exactly sure what something on the screen does. So, you press down with your finger and try dragging the element. It starts to move. Unsure of what will happen, you lift your finger up, and the element moves back into place. You've just discovered how part of this application works.

You'll use the `Scrollable` component to implement swipeable and cancellable behaviors like this. You can create a somewhat generic component that allows the user to swipe text off the screen and, when that happens, call a callback function. Let's look at the code that will render the swipeables before we look at the generic component itself:

```
import React, { useState } from "react";
import { View } from "react-native";
import styles from "./styles";
import Swipeable from "./Swipeable";

export default function SwipableAndCancellable() {
  const [items, setItems] = useState(
    new Array(10).fill(null).map((v, id) => ({ id, name:
      "Swipe Me" }))
  );

  function onSwipe(id) {
    return () => {
      setItems(items.filter((item) => item.id !== id));
    };
  }

  return (
    <View style={styles.container}>
      {items.map((item) => (
        <Swipeable key={item.id} onSwipe={onSwipe(item.id)}
          name={item.name} />
      ))}
    </View>
  );
}
```

This will render 10 `<Swipeable>` components on the screen. Let's see what this looks like:

Figure 24.6 – Screen with Swipeable components

Now, if you start to swipe one of these items to the left, it will move. Here's what this looks like:

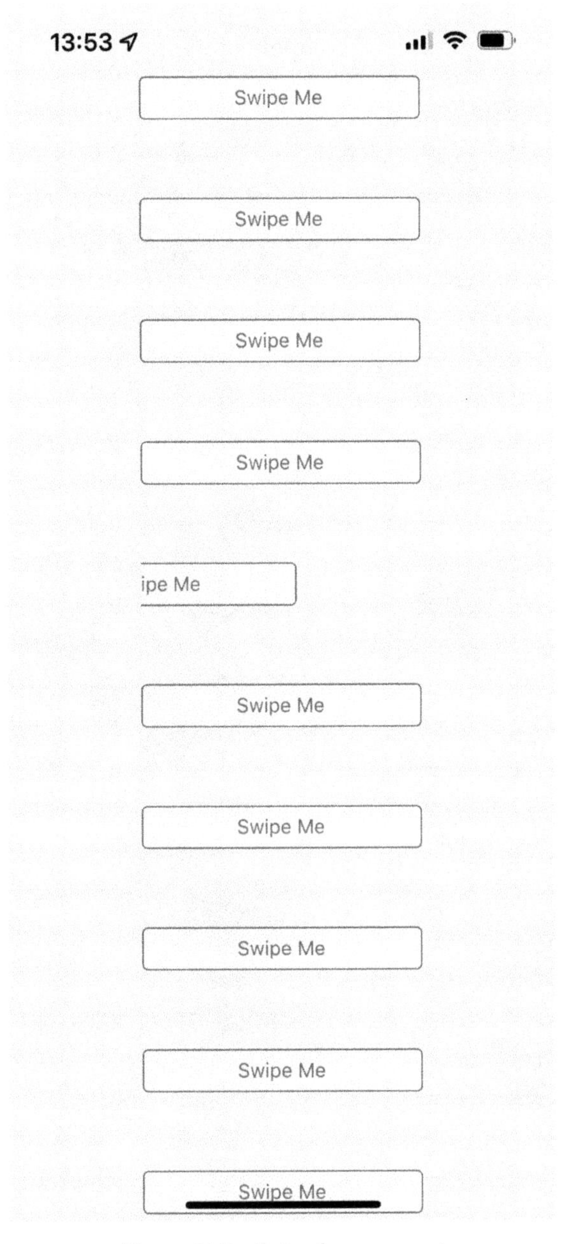

Figure 24.7 – Swiped component

If you don't swipe far enough, the gesture will be canceled and the item will move back into place, as expected. If you swipe it all the way, the item will be removed from the list completely and the items on the screen will fill the empty space.

Now, let's take a look at the `Swipeable` component itself:

```
import React from "react";
import { View, ScrollView, Text, TouchableOpacity } from
  "react-native";
import styles from "./styles";

export default function Swipeable({ onSwipe, name }) {
  function onScroll(e) {
    e.nativeEvent.contentOffset.x === 200 && onSwipe();
  }

  const scrollProps = {
    horizontal: true,
    pagingEnabled: true,
    showsHorizontalScrollIndicator: false,
    scrollEventThrottle: 10,
    onScroll,
  };
```

Here, we have defined a set of props for the `ScrollView` component. Let's apply them to the component and see the layout:

```
  return (
    <View style={styles.swipeContainer}>
      <ScrollView {...scrollProps}>
        <TouchableOpacity>
          <View style={styles.swipeItem}>
            <Text
              style={styles.swipeItemText}>{name}</Text>
          </View>
        </TouchableOpacity>
        <View style={styles.swipeBlank} />
```

```
      </ScrollView>
    </View>
  );
}
```

Note that the `<ScrollView>` component is set to horizontal and that `pagingEnabled` is `true`. It's the paging behavior that snaps the components into place and provides cancellable behavior. This is why there's a blank component besides the component with text in it. Here are the styles that are used for this component:

```
  swipeContainer: {
    flex: 1,
    flexDirection: "row",
    width: 200,
    height: 30,
    marginTop: 50,
  },
  swipeItem: {
    width: 200,
    height: 30,
    backgroundColor: "azure",
    justifyContent: "center",
    borderWidth: 1,
    borderRadius: 4,
    borderColor: "slategrey",
  },
  swipeItemText: {
    textAlign: "center",
    color: "slategrey",
  },
  swipeBlank: {
    width: 200,
    height: 30,
  },
```

The `swipeBlank` style has the same dimensions as `swipeItem`, but nothing else. It's invisible.

We have now covered all the topics in this chapter.

Summary

In this chapter, we were introduced to the idea that gestures on native platforms make a significant difference compared to mobile web platforms. We started off by looking at the `ScrollView` component, and how it makes life much simpler by providing native scrolling behavior for wrapped components.

Next, we spent some time implementing buttons with touch feedback. This is another area that's tricky to get right on the mobile web. We learned how to use the `TouchableOpacity`, `TouchableHighlight`, and `Pressed API` components to do this.

Finally, we implemented a generic `Swipeable` component. Swiping is a common mobile pattern, and it allows the user to discover how things work without feeling intimidated.

In the next chapter, we'll learn how to control animation using React Native.

Further reading

Take a look at the following links for more information:

- `ScrollView`: `https://reactnative.dev/docs/scrollview`
- `TouchableHighlight`: `https://reactnative.dev/docs/touchablehighlight`
- `TouchableOpacity`: `https://reactnative.dev/docs/touchableopacity`
- `Pressable`: `https://reactnative.dev/docs/pressable`

25
Using Animations

Animations can be used to improve the user experience in mobile applications. They usually help users to quickly recognize that something has changed, or help them focus on what is important. Also, animations are simply fun to look at.

There are a couple of different approaches for processing and controlling animations in React Native. Firstly, we will take a look at animation tools that we can use, discover their pros and cons, and compare them. And then we will implement several examples to get to know APIs better.

We'll cover the following topics in this chapter:

- Using React Native Reanimated
- Animating layout components
- Animating styling components

Technical requirements

You can find the code files for this chapter on GitHub at `https://github.com/PacktPublishing/React-and-React-Native-4th-Edition/tree/main/Chapter25`.

Using React Native Reanimated

In the React Native world, we have a lot of libraries and approaches to animate our components, including the built-in Animated API. But in this chapter, I would like to opt for a library called React Native Reanimated and compare it with the Animated API to learn why it is the best choice.

Animated API

The Animated API is the most common tool used to animate components in React Native. It has a set of methods that help you to create an animation object, control its state, and process it. The main benefit is that it can be used with any component, and not just animated components such as `View` or `Text`.

But, at the same time, this API has been implemented in the old architecture of React Native. Asynchronous communications between JavaScript and UI Native threads are used with the Animated API, delaying updates by at least one frame and lasting approximately 16 ms. Sometimes, the delay may last even longer if the JavaScript thread is running React's `diff` algorithm and comparing or processing network requests simultaneously. The problem of dropped or delayed frames can be solved with the React Native Reanimated library, which is based on the new architecture and processes all business logic from the JavaScript thread in the UI thread.

React Native Reanimated

React Native Reanimated can be utilized to provide a more exhaustive abstraction of the Animated API to use with React Native. It provides an imperative API with multistage animations and custom transitions, while at the same time providing a declarative API that can be used to describe simple animations and transitions in a similar way to how CSS transitions work. It's built on top of React Native Animated and reimplements it on the Native Thread. This allows you to use the familiar JavaScript language while taking advantage of the highest performance and simplest API.

Furthermore, React Native Reanimated defines worklets, which are JavaScript functions that can be synchronously executed within the UI thread. This allows instant animations without having to wait for a new frame. Let's take a look at what a simple worklet looks like:

```
function simpleWorklet() {
  "worklet";
  console.log("Hello from UI thread");
}
```

The only thing that is needed for the `simpleWorklet` function to get called inside the UI thread is to add the `"worklet"` directive at the top of the function block.

React Native Reanimated provides a variety of hooks and methods that help us to handle animations:

- `useSharedValue`: This hook returns a `SharedValue` instance, which is the main stateful data object that lives in the UI thread context and has a similar concept to `Animated.Value` in the core Animated API. A **Reanimated** animation is triggered when `SharedValue` is changed. This is similar to how React re-rendered components when the state was changed.
- `useDerivedValue`: This allows the creation of another value based on a `SharedValue`.
- `useAnimatedStyle`: This hook creates an animated, observed style object in React Native.
- `withTiming`, `withSpring`, `withDecay`: These are utility methods that update `SharedValue` smoothly with animation.

We have learned what React Native Reanimated is and how it is different from the Animated API. Next, let's try to install it and apply it to our app.

Installing the React Native Reanimated library

To install the React Native Reanimated library, run this command inside your Expo project:

```
expo install react-native-reanimated
```

After the installation is complete, add the Babel plugin to `babel.config.js`:

```
module.exports = function(api) {
  api.cache(true);
  return {
    presets: ['babel-preset-expo'],
    plugins: ['react-native-reanimated/plugin'],
  };
};
```

After you add the Babel plugin, restart your development server and clear the bundler cache:

```
expo start --clear
```

This section has introduced us to the React Native Reanimated library. We have found out why it is better than the built-in Animated API. In the next sections, we will use it in real examples.

Animating layout components

A common use case is animating the entering and exiting layout of your components. This means that when your component renders for the first time and when you unmount your component, it appears animated. React Native Reanimated is an API that lets you animate layouts and add animations such as `FadeIn`, `BounceIn`, and `ZoomIn`.

React Native Reanimated also provides a special `Animated` component that is the same as the `Animated` component in the Animated API, but with additional props:

- `entering`: Accepts predefined animation when the component will mount and render
- `exiting`: Accepts the same animation object, but it will be called when the component unmounts

Let's create a simple todo list with a button for creating tasks, and a feature that allows us to delete tasks when we click on them.

> **Important Note**
>
> It's impossible to see animation in screenshots, so I suggest you open the code and try to implement the animations to see the results.

Firstly, let's take a look at the main screen of our todo list app and how the items are rendering at the moment:

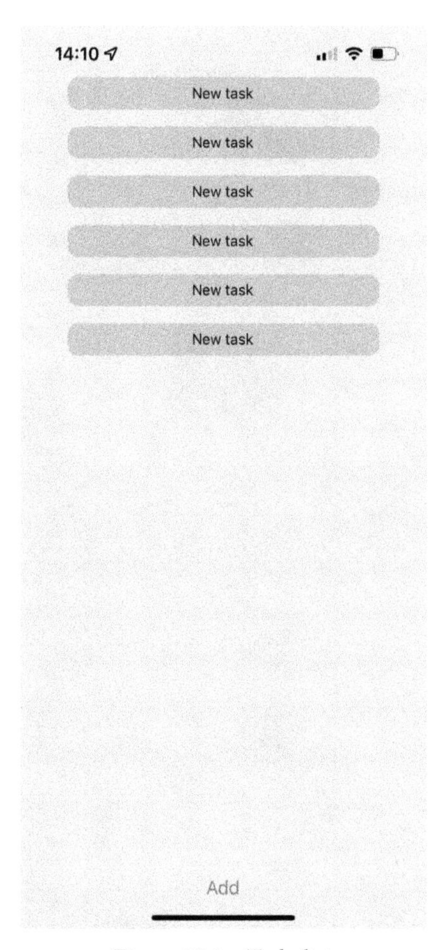

Figure 25.1 – Todo list

This is a simple example with a list of task items and one button for adding new tasks. When we quickly press the **Add** button several times, the list items come from the left side of the screen with an animation:

Figure 25.2 – Todo list with animated rendering

The magic is implemented in the `TodoItem` component. Let's take a look at it:

```
import { Text, TouchableOpacity } from "react-native";
import Animated, { SlideInLeft, SlideOutRight } from
  "react-native-reanimated";
import { styles } from "./styles";

export const TodoItem = ({ id, title, onPress }) => {
  return (
    <Animated.View entering={SlideInLeft}
      exiting={SlideOutRight}>
      <TouchableOpacity onPress={() => onPress(id)}
        style={styles.todoItem}>
        <Text>{title}</Text>
      </TouchableOpacity>
    </Animated.View>
  );
};
```

As you can see, there is no complicated logic, and there's not too much code. We just take the `Animated` component as the root of animation and pass predefined animations from the React Native Reanimated library to the `entering` and `exiting` props.

To see how the items disappear from the screen, we need to press the todo items so the exiting animation will run. I've pressed a few items and tried to catch the result in the following screenshot:

Figure 25.3 – Deleting todo items from the screen

Let's examine the App component to see the entire picture:

```
export default function App() {
  const [todoList, setTodoList] = useState([]);
  const addTask = () => {
    setTodoList([
      ...todoList,
      { id: String(new Date().getTime()), title: "New task"
        },
    ]);
  };
  const deleteTask = (id) => {
    setTodoList(todoList.filter((todo) => todo.id !== id));
  };
```

We have created a todoList state using the useState hook and handler functions for adding and deleting tasks. Next, let's take a look at how the animation will be applied to the layout:

```
  return (
    <View style={styles.container}>
      <View style={{ flex: 1 }}>
        {todoList.map(({ id, title }) => (
          <TodoItem key={id} id={id} title={title}
            onPress={deleteTask} />
        ))}
      </View>
      <Button onPress={addTask} title="Add" />
    </View>
  );
}
```

In this example, we learned a simple way to apply animations to make our app look better. However, the React Native Reanimated library is a lot more powerful than we imagined. The next example illustrates how we can animate and create our own animations by applying them directly to the styles of our components.

Animating styling components

In a more complex example, I suggest creating a button with beautiful tappable feedback. This button will be built using the `Pressable` component that we learned about in *Chapter 24, Responding to User Gestures*. This component accepts the `onPressIn`, `onLongPress`, and `onPressOut` events. As a result of these events, we will be able to see how our touches will be reflected on the button.

Let's start by defining `SharedValue` and `AnimatedStyle`:

```
const radius = useSharedValue(30);
const opacity = useSharedValue(1);
const scale = useSharedValue(1);
const color = useSharedValue(0);

const animatedStyles = useAnimatedStyle(() => {
  const backgroundColor = interpolateColor(
    color.value,
    [0, 1],
    ["orange", "red"]
  );

  return {
    opacity: opacity.value,
    borderRadius: radius.value,
    transform: [{ scale: scale.value }],
    backgroundColor: backgroundColor,
  };
}, []);
```

In order to animate style properties, we have created a `SharedValue` object using the `useSharedValue` hook. It takes default values as an argument. Next, we created the style object with the `useAnimatedStyle` hook. The hook accepts the callback that should return a style object. The `useAnimatedStyle` hook is similar to the `useMemo` hook, but all calculations are performed in the UI thread and all `SharedValue` changes will invoke the hook to recalculate the style object.

Next, let's create handler functions that will update the style properties in relation to the pressing state of the button:

```
const onPressIn = () => {
  radius.value = withSpring(20);
  opacity.value = withSpring(0.7);
  scale.value = withSpring(0.9);
};

const onLongPress = () => {
  scale.value = withSpring(0.8);
  color.value = withSpring(1);
};

const onPressOut = () => {
  radius.value = withSpring(30);
  opacity.value = withSpring(1);
  scale.value = withSpring(1, { damping: 50 });
  color.value = withSpring(0);
};
```

The first handler, `onPressIn`, updates `borderRadius`, `opacity`, and `scale` from their default values. We also update these values using `withSpring`, which makes updating styles smoother. Like the first handler, other ones will also update the style of the button but in different ways. `onLongPress` turns the button red and makes it smaller. `onPressOut` resets all values to their default values.

We've implemented all necessary logic and can now apply it to the layout:

```
<View style={styles.container}>
  <Animated.View style={[styles.buttonContainer,
    animatedStyles]}>
    <Pressable
      onPressIn={onPressIn}
      onPressOut={onPressOut}
      onLongPress={onLongPress}
      style={styles.button}
    >
```

```
            <Text style={styles.buttonText}>Press me</Text>
          </Pressable>
        </Animated.View>
      </View>
```

Finally, let's take a look at the result:

Figure 25.4 – Button with default, pressed, and long-pressed styles

In *Figure 25.4*, you can see the three states of the button: default, pressed, and long-pressed.

Summary

In this chapter, we've learned how to use the React Native Reanimated library to add animations to the layout and components. We've gone through the basic principles of the library and found out how it works under the hood and how it executes code inside the UI thread without using `Bridge` to connect JavaScript and Native layers of the app.

We also went through two examples using the React Native Reanimated library. In the first example, we learned how to apply a layout animation using predefined declarative animations to get our component to appear and disappear beautifully. In the second example, we animated the button's styles with the `useSharedValue` and `useAnimatedStyle` hooks.

Skills to animate components and layout will help you make your app more beautiful and responsive. In the next chapter, we'll learn about controlling images in our apps.

Further reading

Check out the following link for more information:

- React Native Reanimated: `https://docs.swmansion.com/react-native-reanimated/docs`

26 Controlling Image Display

So far, the examples in this book haven't rendered any images on mobile screens. This doesn't reflect the reality of mobile applications. Web applications display lots of images. If anything, native mobile applications rely on images even more than web applications because images are a powerful tool when you have a limited amount of space.

In this chapter, you'll learn how to use the React Native `Image` component, starting with loading images from different sources. Then, you'll learn how you can use the `Image` component to resize images, and how you can set placeholders for lazily loaded images. Finally, you'll learn how to implement icons using the `@expo/vector-icons` package. These sections cover the most common use cases for using images and icons in apps.

We'll cover the following topics in this chapter:

- Loading images
- Resizing images
- Lazy image loading
- Rendering icons

Technical requirements

You can find the code files for this chapter on GitHub at `https://github.com/PacktPublishing/React-and-React-Native-4th-Edition/tree/main/Chapter26`.

Loading images

Let's get started by figuring out how to load images. You can render the `<Image>` component and pass its properties just like any other React component. But this particular component needs image blob data to be of any use. Let's look at some code:

```
export default function App({ reactSource, relaySource }) {
  return (
    <View style={styles.container}>
      <Image style={styles.image} source={reactSource} />
      <Image style={styles.image} source={relaySource} />
    </View>
  );
}
const sourceProp = PropTypes.oneOfType([
  PropTypes.shape({
    uri: PropTypes.string.isRequired,
  }),
  PropTypes.number,
]).isRequired;
App.propTypes = {
  reactSource: sourceProp,
  relaySource: sourceProp,
};
App.defaultProps = {
  reactSource: {
    uri: "https://reactnative.dev/docs/assets/favicon.png",
  },
  relaySource: require("./images/relay.png"),
};
```

There are two ways to load the blob data into an `<Image>` component. The first approach loads the image data from the network. This is done by passing an object with a `uri` property to `source`. The second `<Image>` component in this example is using a local image file. It does this by calling `require()` and passing the result to `source`.

Take a look at the `sourceProp` property type validator. This gives you an idea of what can be passed to the `source` property. It's either an object with a `uri` string property or a number. It expects a number because `require()` returns a number.

Now, let's see what the rendered result looks like:

Figure 26.1 – Image loading

Here's the style that was used with these images:

```
image: {
  width: 100,
  height: 100,
  margin: 20,
},
```

Note that without the `width` and `height` style properties, images will not render. In the next section, you'll learn how image resizing works when the `width` and `height` values are set.

Resizing images

The `width` and `height` style properties of `Image` components determine the size of what's rendered on the screen. For example, you'll probably have to work with images at some point that have a larger resolution than you want to be displayed in your React Native application. Simply setting the `width` and `height` style properties on the `Image` is enough to properly scale the image.

Let's look at some code that lets you dynamically adjust the dimensions of an image using controls:

```
import React, { useState } from "react";
import { View, Text, Image } from "react-native";
import Slider from "@react-native-community/slider";
import styles from "./styles";

export default function App() {
  const source = require("./images/flux.png");
  const [width, setWidth] = useState(100);
  const [height, setHeight] = useState(100);

  return (
    <View style={styles.container}>
      <Image source={source} style={{ width, height }} />
      <Text>Width: {width}</Text>
      <Text>Height: {height}</Text>
      <Slider
```

```
          style={styles.slider}
          minimumValue={50}
          maximumValue={150}
          value={width}
          onValueChange={(value) => {
            setWidth(value);
            setHeight(value);
          }}
        />
      </View>
    );
  }
```

Here's what the image looks like if you're using the default 100 x 100 dimensions:

Figure 26.2 – Image 100 x 100

Here's a scaled-down version of the image:

Figure 26.3 – Image 50 x 50

Lastly, here's a scaled-up version of the image:

Figure 26.4 – Image 150 x 150

Important Note

There's a `resizeMode` property that you can pass to `Image` components. This determines how the scaled image fits within the dimensions of the actual component. You'll see this property in action in the *Rendering icons* section of this chapter.

As you can see, the dimensions of the images are controlled by the `width` and `height` style properties. Images can even be resized while the app is running by changing these values. In the next section, you'll learn how to lazily load images.

Lazy image loading

Sometimes, you don't necessarily want an image to load at the exact moment that it's rendered; for example, you might be rendering something that's not visible on the screen yet. Most of the time, it's perfectly fine to fetch the image source from the network before it's actually visible. But if you're fine-tuning your application and discover that loading lots of images over the network causes performance issues, you can lazily load the source.

I think the more common use case in a mobile context is handling a scenario where you've rendered one or more images where they're visible, but the network is slow to respond. In this case, you will probably want to render a placeholder image so that the user sees something right away, rather than an empty space. So, let's get started:

1. Firstly, you can implement an abstraction that wraps the actual image that you want to show once it's loaded. Here's the code for this:

```
import React, { useState } from "react";
import PropTypes from "prop-types";
import { View, Image } from "react-native";

const placeholder =
  require("./assets/placeholder.png");

function Placeholder(props) {
  if (props.loaded) {
    return null;
  } else {
    return <Image style={props.style}
      source={placeholder} />;
  }
}
```

Now, here, you can see the placeholder image will be rendered only while the original image isn't loaded:

```
export default function LazyImage(props) {
  const [loaded, setLoaded] = useState(false);

  return (
    <View style={props.style}>
      <Placeholder loaded={loaded} {...props} />
      <Image
        {...props}
        onLoad={() => {
          setLoaded(true);
        }}
      />
    </View>
  );
}

LazyImage.propTypes = {
  style: PropTypes.shape({
    width: PropTypes.number.isRequired,
    height: PropTypes.number.isRequired,
  }),
};
```

This component renders a `View` component with two `Image` components inside it. It also has a `loaded` state, which is initially `false`. When `loaded` is `false`, the placeholder image is rendered. The `loaded` state is set to `true` when the `onLoad()` handler is called. This means that the placeholder image is removed and the main image is displayed.

2. Now, let's use the `LazyImage` component that we've just implemented. You'll render the image without a source, and the placeholder image should be displayed. Let's add a button that gives the lazy image a source. When it loads, the placeholder image should be replaced. Here's what the main app module looks like:

```
import React, { useState } from "react";
import { View } from "react-native";
import styles from "./styles";
import LazyImage from "./LazyImage";
import Button from "./Button";

const remote =
  "https://reactnative.dev/docs/assets/favicon.png";

export default function LazyLoading() {
  const [source, setSource] = useState(null);

  return (
    <View style={styles.container}>
      <LazyImage
        style={{ width: 200, height: 150 }}
        resizeMode="contain"
        source={source}
      />
      <Button
        label="Load Remote"
        onPress={() => {
          setSource({ uri: remote });
        }}
      />
    </View>
  );
}
```

This is what the screen looks like initially:

Figure 26.5 – Initial state of the image

3. Then, click the **Load Remote** button to eventually see the image that we actually want:

Figure 26.6 – Loaded image

You might notice that, depending on your network speed, the placeholder image remains visible, even after you click the **Load Remote** button. This is by design because you don't want to remove the placeholder image until you know for sure that the actual image is ready to be displayed. Now, let's render some icons in our React Native application.

Rendering icons

In the final section of this chapter, you'll learn how to render icons in React Native components. Using icons to indicate meaning makes web applications more usable. So, why should native mobile applications be any different?

We'll use the `@expo/vector-icons` package to pull various vector font packages into your React Native app. This package is already part of the Expo project that we use as base of the app and now, you can import `Icon` components and render them. Let's implement an example that renders several `FontAwesome` icons based on a selected icon category:

```
import React, { useState, useEffect } from "react";
import { View, FlatList, Text } from "react-native";
import { Picker } from "@react-native-picker/picker";
import Icon from "@expo/vector-icons/FontAwesome";
import styles from "./styles";
import iconNames from "./icon-names.json";

export default function RenderingIcons() {
  const [selected, setSelected] = useState("Web Application
    Icons");
  const [listSource, setListSource] = useState([]);
  const categories = Object.keys(iconNames);

  function updateListSource(selected) {
    setListSource(iconNames[selected]);
    setSelected(selected);
  }

  useEffect(() => {
    updateListSource(selected);
  }, []);
```

Here, we have defined all necessary logic to store and update the icon data. Next, we will apply it to the layout:

```
  return (
    <View style={styles.container}>
      <View style={styles.picker}>
        <Picker selectedValue={selected}
          onValueChange={updateListSource}>
          {categories.map((category) => (
            <Picker.Item key={category} label={category}
              value={category} />
          ))}
        </Picker>
      </View>
      <FlatList
        style={styles.icons}
        data={listSource.map((value, key) => ({ key:
          key.toString(), value }))}
        renderItem={({ item }) => (
          <View style={styles.item}>
            <Icon name={item.value} style={styles.itemIcon}
              />
            <Text
              style={styles.itemText}>{item.value}</Text>
          </View>
        )}
      />
    </View>
  );
}
```

When you run this example, you should see something that looks like the following:

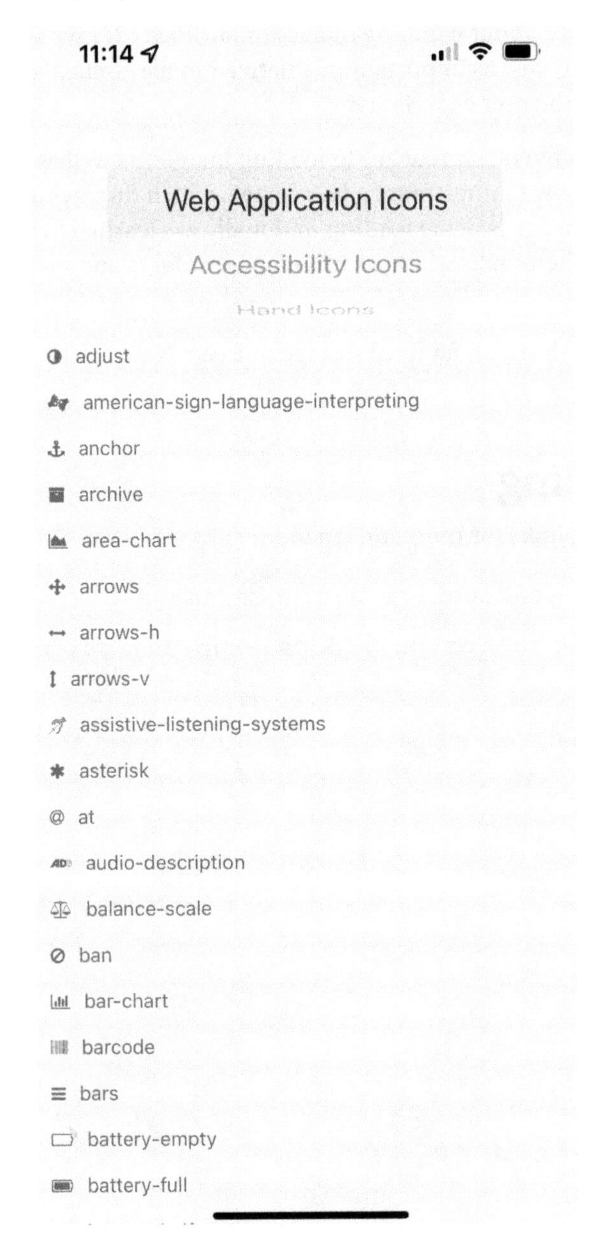

Figure 26.7 – Rendering icons

The color of each icon is specified in the same way you would specify the color of text: via styles.

Summary

In this chapter, we learned about handling images in our React Native applications. Images in a native application are just as important in a native mobile context as they are in a web context – they improve the user experience.

We learned about the different approaches to loading images, as well as how to resize them. We also learned how to implement a lazy image, which displays a placeholder image while the actual image is being loading in. Finally, we learned how to use icons in a React Native app. These skills will help you manage images and make your app more informative.

In the next chapter, we'll learn about local storage in React Native, which is handy when our app goes offline.

Further reading

Check out the following links for more information:

- Image: `https://reactnative.dev/docs/image`
- React Native vector icons: `https://docs.expo.dev/guides/icons/`

27
Going Offline

Users expect applications to operate seamlessly with unreliable network connections. If your mobile application can't cope with transient network issues, your users will use a different app. When there's no network, you have to persist data locally on the device. Alternatively, perhaps your app doesn't even require network access, in which case you'll still need to store data locally.

In this chapter, you'll learn how to do the three things with React Native. First, you'll learn how to detect the state of the network connection. Second, you'll learn how to store data locally. Lastly, you'll learn how to synchronize local data that's been stored due to network problems once it comes back online.

In this chapter, we'll cover the following topics:

- Detecting the state of the network
- Storing application data
- Synchronizing application data

Technical requirements

You can find the code files for this chapter on GitHub at `https://github.com/PacktPublishing/React-and-React-Native-4th-Edition/tree/main/Chapter27`.

Detecting the state of the network

If your code tries to make a request over the network while disconnected – using `fetch()`, for example – an error will occur. You probably have error-handling code in place for these scenarios already, since the server could return some other type of error.

However, in the case of connectivity trouble, you might want to detect this issue before the user attempts to make network requests.

There are two potential reasons for proactively detecting the network state. The first one is to prevent the user from performing any network requests until you've detected that the app is back online. To do that, you can display a friendly message to the user stating that, since the network is disconnected, they can't do anything. The other possible benefit of early network state detection is that you can prepare to perform actions offline and sync the app state when the network is connected again.

Let's look at some code that uses the `NetInfo` utility from the `@react-native-community/netinfo` package to handle changes in network state:

```
import React, { useState, useEffect } from "react";
import { Text, View } from "react-native";
import NetInfo from "@react-native-community/netinfo";
import styles from "./styles";

const connectedMap = {
  none: "Disconnected",
  unknown: "Disconnected",
  wifi: "Connected",
  cell: "Connected",
  mobile: "Connected",
};
```

`connectedMap` covers all connection states and will help us to render it on the screen. Let's now see the `App` component:

```
export default function App() {
  const [connected, setConnected] = useState("");

  useEffect(() => {
    function onNetworkChange(connection) {
      setConnected(connectedMap[connection.type]);
    }

    const unsubscribe =
      NetInfo.addEventListener(onNetworkChange);

    return () => {
      unsubscribe();
    };
  }, []);

  return (
    <View style={styles.container}>
      <Text>{connected}</Text>
    </View>
  );
}
```

This component will render the state of the network based on the string values in `connectedMap`. The `connectionChange` event of the `NetInfo` object will cause the `connected` state to change. For example, when you run this app for the first time, the screen might look like this:

Figure 27.1 – Connected state

Then, if you turn off networking on your host machine, the network state will change on the emulated device as well, causing the state of our application to change, as follows:

Figure 27.2 – Disconnected state

In the next section, you'll learn how to store application data locally on the device where the application is running.

Storing application data

The `AsyncStorage` API works the same on both the iOS and Android platforms. You would use this API for applications that don't require any network connectivity in the first place or to store data that will eventually be synchronized using an API endpoint once a network becomes available.

To install the `async-storage` package, run the following command:

```
expo install @react-native-async-storage/async-storage
```

Let's look at some code that allows the user to enter a key and a value and then stores them:

```
import React, { useState, useEffect } from "react";
import { Text, TextInput, View, FlatList } from
  "react-native";
import AsyncStorage from
  "@react-native-async-storage/async-storage";
import styles from "./styles";
import Button from "./Button";

export default function App() {
  const [key, setKey] = useState(null);
  const [value, setValue] = useState(null);
  const [source, setSource] = useState([]);
```

The `key`, `value`, and `source` values will handle our state. To save it in `AsyncStorage`, we need to define functions:

```
  function setItem() {
    return AsyncStorage.setItem(key, value)
      .then(() => {
        setKey(null);
        setValue(null);
      })
      .then(loadItems);
  }
```

```
function clearItems() {
  return AsyncStorage.clear().then(loadItems);
}

async function loadItems() {
  const keys = await AsyncStorage.getAllKeys();
  const values = await AsyncStorage.multiGet(keys);
  setValues(values);
}

useEffect(() => {
  loadItems();
}, []);
```

Here's the markup that's rendered by the `App` component:

```
return (
  <View style={styles.container}>
    <Text>Key:</Text>
    <TextInput
      style={styles.input}
      value={key}
      onChangeText={(v) => {
        setKey(v);
      }}
    />
    <Text>Value:</Text>
    <TextInput
      style={styles.input}
      value={value}
      onChangeText={(v) => {
        setValue(v);
      }}
```

```
      />
      <View style={styles.controls}>
        <Button label="Add" onPress={setItem} />
        <Button label="Clear" onPress={clearItems} />
      </View>
```

The markup in the preceding code block is represented as inputs and buttons to create, save, and delete items. Next, we will render the list of items by the `FlatList` component:

```
      <View style={styles.list}>
        <FlatList
          data={source.map(([key, value]) => ({
            key: key.toString(),
            value,
          }))}
          renderItem={({ item: { value, key } }) => (
            <Text>
              {value} ({key})
            </Text>
          )}
        />
      </View>
    </View>
  );
```

Before we walk through what this code is doing, let's take a look at the following screen, since it'll explain most of what we're going to cover under storing application data:

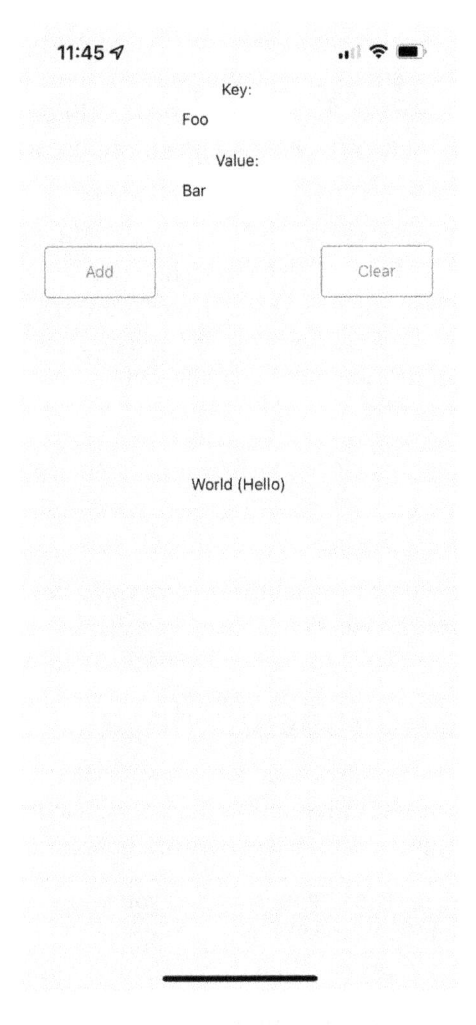

Figure 27.3 – Storing application data

As you can see in *Figure 27.3*, there are two input fields and two buttons. The fields allow the user to enter a new key and value. The **Add** button allows the user to store this key-value pair locally on their device, while the **Clear** button clears any existing items that have been stored previously.

The AsyncStorage API works the same for both iOS and Android. Under the hood, AsyncStorage works very differently, depending on which platform it's running on. The reason React Native is able to expose the same storage API on both platforms is due to its simplicity – it's just key-value pairs. Anything more complex than that is left up to the application developer.

The abstractions that you've created around AsyncStorage in this example are minimal. The idea is to set and get items. However, even straightforward actions like this deserve an abstraction layer. For example, the setItem() method you've implemented here will make the asynchronous call to AsyncStorage and update the items state once that has been completed. Loading items is even more complicated because you need to get the keys and values as two separate asynchronous operations.

The reason we do this is to keep the UI responsive. If there are pending screen repaints that need to happen while data is being written to disk, preventing those from happening by blocking them would lead to a suboptimal user experience.

In the next section, you'll learn how to synchronize data that's been stored locally while the device is offline with remote services once the device comes back online.

Synchronizing application data

So far in this chapter, you've learned how to detect the state of a network connection and how to store data locally in a React Native application. Now, it's time to combine these two concepts and implement an app that can detect network outages and continue to function.

The basic idea is to only make network requests when you know for sure that the device is online. If you know that it isn't, you can store any changes in the state locally. Then, when you're back online, you can synchronize those stored changes with the remote API.

Let's implement a simplified React Native app that does this. The first step is to implement an abstraction that sits between the React components and the network calls that store data. We'll call this module store.js:

```
export function set(key, value) {
  return new Promise((resolve, reject) => {
    if (connected) {
      fakeNetworkData[key] = value;
      resolve(true);
    } else {
      AsyncStorage.setItem(key, value.toString()).then(
```

```
          () => {
            unsynced.push(key);
            resolve(false);
          },
          (err) => reject(err)
        );
      }
    });
  }
```

The `set` method depends on the `connected` variable, and depending on whether there is an internet connection or not, it handles the different logic. Actually, the `get` method also follows the same approach:

```
export function get(key) {
  return new Promise((resolve, reject) => {
    if (connected) {
      resolve(key ? fakeNetworkData[key] :
        fakeNetworkData);
    } else if (key) {
      AsyncStorage.getItem(key).then(
        (item) => resolve(item),
        (err) => reject(err)
      );
    } else {
      AsyncStorage.getAllKeys().then(
        (keys) =>
          AsyncStorage.multiGet(keys).then(
            (items) => resolve(Object.fromEntries(items)),
            (err) => reject(err)
          ),
        (err) => reject(err)
      );
    }
  });
}
```

This module exports two functions – `set()` and `get()`. Their jobs are to set and get data, respectively. Since this is just a demonstration of how to sync between local storage and network endpoints, this module just mocks the actual network with the `fakeNetworkData` object.

Let's start by looking at the `set()` function. It's an asynchronous function that will always return a promise that resolves to a Boolean value. If it's `true`, it means that you're online and that the call over the network was successful. If it's `false`, it means that you're offline, and `AsyncStorage` was used to save the data.

The same approach is used with the `get()` function. It returns a promise that resolves a Boolean value that indicates the state of the network. If a key argument is provided, then the value for that key is looked up. Otherwise, all the values are returned, either from the network or from `AsyncStorage`.

In addition to these two functions, this module does two other things. It uses `NetInfo.fetch()` to set the `connected` state. Then, it adds a listener to listen for changes in the network state. This is how items that were saved locally when you were offline become synced with the network when it's connected again.

Now, let's check out the main application that uses these functions:

```
export default function App() {
  const [message, setMessage] = useState(null);
  const [first, setFirst] = useState(false);
  const [second, setSecond] = useState(false);
  const [third, setThird] = useState(false);
  const setters = new Map([
    ["first", setFirst],
    ["second", setSecond],
    ["third", setThird],
  ]);
```

Here, we have defined the `state` variables that we will use in the `Switch` components:

```
function save(key) {
  return (value) => {
    set(key, value).then(
      (connected) => {
        setters.get(key)(value);
        setMessage(connected ? null : "Saved Offline");
      },
      (err) => {
        setMessage(err);
      }
    );
  };
}
```

The `save()` function helps us to reuse logic in a different `Switch` component. Next, we have the `useEffect` hook to get saved data when the page renders for the first time:

```
useEffect(() => {
  NetInfo.fetch().then(() =>
    get().then(
      (items) => {
        for (let [key, value] of Object.entries(items)) {
          setters.get(key)(value);
        }
      },
      (err) => {
        setMessage(err);
      }
    )
  );
}, []);
```

Next, let's take a look at the final markup of the page:

```
  return (
    <View style={styles.container}>
      <Text>{message}</Text>
      <View>
        <Text>First</Text>
        <Switch
          value={boolMap[first.toString()]}
          onValueChange={save("first")}
        />
      </View>
      <View>
        <Text>Second</Text>
        <Switch
          value={boolMap[second.toString()]}
          onValueChange={save("second")}
        />
      </View>
      <View>
        <Text>Third</Text>
        <Switch
          value={boolMap[third.toString()]}
          onValueChange={save("third")}
        />
      </View>
    </View>
  );
}
```

The job of the `App` component is to save the state of three `Switch` components, which is difficult when you're providing the user with a seamless transition between online and offline modes. Thankfully, your `set()` and `get()` abstractions, which are implemented in another module, hide most of the details from the application's functionality.

Note, however, that you need to check the state of the network in this module before you attempt to load any items. If you don't do this, then the `get()` function will assume that you're offline, even if the connection is fine. Here's what the app looks like:

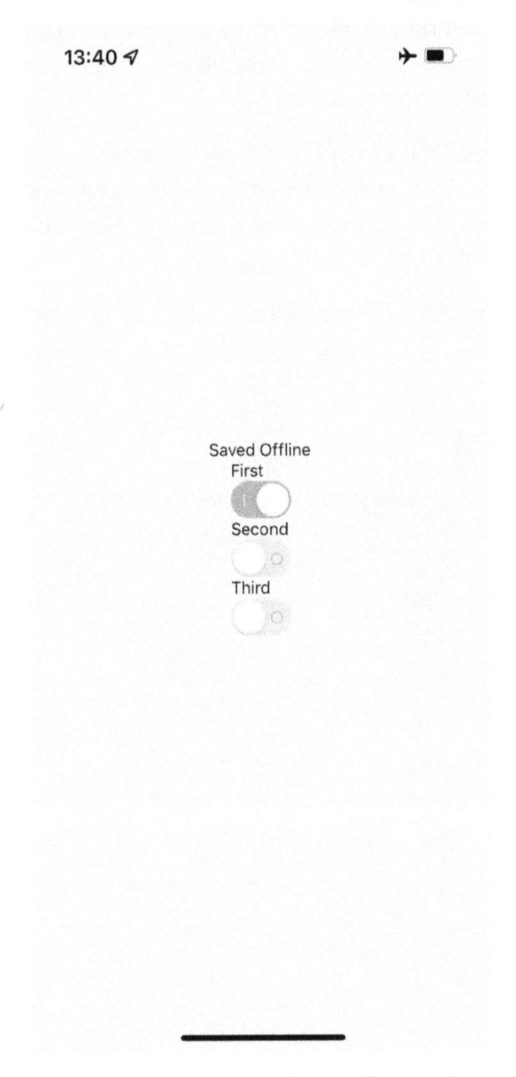

Figure 27.4 – Synchronizing application data

Note that you won't actually see the **Saved Offline** message until you change something in the UI.

Summary

This chapter introduced us to storing data offline in React Native applications. The main reason we would want to store data locally is when the device goes offline and our app can't communicate with a remote API. However, not all applications require API calls, and `AsyncStorage` can be used as a general-purpose storage mechanism. We just need to implement the appropriate abstractions around it.

We also learned how to detect changes in the network state of React Native apps. It's important to know when the device has gone offline so that our storage layer doesn't make pointless attempts at network calls. Instead, we can let the user know that the device is offline and then synchronize the application state when a connection is available.

In the next chapter, we'll learn how to import and use UI components from the `NativeBase` library.

Further reading

You can find more information on `AsyncStorage` at `https://react-native-async-storage.github.io/async-storage/`.

28

Selecting Native UI Components Using NativeBase

Right out of the box, React Native gives us most of the tools we need to build a fully functional native application that runs on both Android and iOS. However, taking your application to the next level and delivering a consistent and polished **user experience** (**UX**) across both platforms requires help. NativeBase can provide us with additional tools that can facilitate quality **user interface** (**UI**) designs for React Native apps. It is possible to build a quality native UI without a tool such as NativeBase, but this would require a lot more coding on our part. If you want to deliver applications that address specific challenges faced by your users rather than maintaining your own UI library, NativeBase might be what you're looking for.

We'll cover the following topics in this chapter:

- Application containers
- Headers, footers, and navigation
- Using layout components
- Collecting input using form components

Technical requirements

You can find the code files for this chapter on GitHub at `https://github.com/PacktPublishing/React-and-React-Native-4th-Edition/tree/main/Chapter28`.

Application containers

Before we can pull out the NativeBase UI components and render them on our application screens, there are a couple of initialization tasks we have to perform. NativeBase requires you to load font files in order to work. Additionally, we want to set up the same general screen structure for every screen using top-level `NativeBase` components. To accomplish both of these goals, we can implement a `Container` component, which can then be used by our screens. We'll start by importing everything that we need:

```
import React from "react";
import { StatusBar } from "react-native";
import { NativeBaseProvider, Box, Text, HStack } from
  "native-base";
import AppLoading from "expo-app-loading";
import {
  useFonts,
  Roboto_500Medium,
  Roboto_400Regular,
} from "@expo-google-fonts/roboto";
import { Ionicons } from "@expo/vector-icons";
import { theme } from "./theme";
```

Now, we can implement the `Container` component:

```
export default function Container({ title, children }) {
  const [fontsLoaded] = useFonts({
    Roboto_500Medium,
    Roboto_400Regular,
    ...Ionicons.font,
  });
```

The `useFonts` Hook helps us to load fonts. When they are loaded, we will render the main content of the page:

```
  if (fontsLoaded) {
    return (
      <NativeBaseProvider theme={theme}>
        <StatusBar bg="#3700B3" barStyle="light-content" />
        <Box safeAreaTop bg="#6200ee" />
        <HStack
          bg="#6200ee"
          px="1"
          py="3"
          justifyContent="center"
          alignItems="center"
          w="100%"
        >
          <Text color="white" fontSize="20"
            fontWeight="bold">
            {title}
          </Text>
        </HStack>
        <Box>{children}</Box>
      </NativeBaseProvider>
    );
  } else {
    return <AppLoading />;
  }
}
```

Let's take a look at what this code is doing. The `Container` component accepts two properties: `title` and `children`. The `title` property is a string that sets the title for each screen in the app, while `children` signifies the contents for each page in the app. The `ready` state is used to determine whether the fonts that we need to load have finished loading or not. The `useFonts()` Hook has been used to load the fonts that NativeBase requires. Without these fonts, we can't render the components. This is why we check the `fontsLoaded` state. If it's `false`, we render the `AppLoading` component while the fonts load. When they finish loading, it sets the `fontsLoaded` state to `true` and renders the `NativeBaseProvider` component.

NativeBase has a `NativeBaseProvider` component, which should be placed at the root of the app. Inside `NativeBaseProvider`, we have a `Header` component implementation based on the `HStack` component. The header is also where we set the title. The contents of a given page are passed in via the `children` property and this value is rendered in the `Box` component.

Let's see how the `Container` component is used by the `App` component:

```
import React from "react";
import { Text } from "native-base";
import Container from "./Container";

export default function App() {
  return (
    <Container title="Application Container">
      <Text>Application content goes here...</Text>
    </Container>
  );
}
```

The title "`Application Container`" is passed as a property of `Container` and the application content is written in simple text, for now, set using the `Text` component. Here's what the result looks like on the screen:

Figure 28.1 – Application container

Now that we have a `Container` component that we can use on every page of the app, let's add in some navigation capabilities and footer navigation links. The approach to creating such containers helps us to reuse a lot of code and manage one `Container` component by passing necessary props to it.

Headers and footers

To implement navigation in our app, we can use the `react-navigation` package that we learned about in *Chapter 18, Navigating between Screens*. But, in this example, we will implement the layout of tab navigation without navigation features.

Here's what our `App` component looks like:

```
import React from "react";
import { Text } from "native-base";
import Container from "./Container";

export default function App() {
  return (
    <Container title="Home">
      <Text>Home content goes here...</Text>
    </Container>
  );
}
```

We're passing to the `Container` component the title that allows us to update the `Header` title. Let's take a look at the updated `Container` component:

```
<NativeBaseProvider theme={theme}>
  <StatusBar bg="#3700B3" barStyle="light-content" />
  <Box safeAreaTop bg="#6200ee" />
  <HStack
    bg="#6200ee"
    px="1"
    py="3"
    justifyContent="center"
    alignItems="center"
    w="100%"
  >
```

```
    <Text color="white" fontSize="20" fontWeight="bold">
      {title}
    </Text>
  </HStack>

  <Box flex={1} safeAreaTop width="100%"
    alignSelf="center">
    <Center flex={1}>
      <Box>{children}</Box>
    </Center>
```

Here, we can see the header and main content of the page. Next, let's add a bottom tab bar:

```
    <HStack bg="#6200ee" alignItems="center" safeAreaBottom
      shadow={6}>
      <FooterButton
        title="Home"
        iconName="home"
        selected={selected === 0}
        onPress={() => setSelected(0)}
      />

      <FooterButton
        title="Settings"
        iconName="settings"
        selected={selected === 1}
        onPress={() => setSelected(1)}
      />

      <FooterButton
        title="Help"
        iconName="help"
        selected={selected === 2}
        onPress={() => setSelected(2)}
      />

      <FooterButton
```

```
          title="Contact"
          iconName="person"
          selected={selected === 3}
          onPress={() => setSelected(3)}
        />
      </HStack>
    </Box>
  </NativeBaseProvider>
```

We've added the `FooterButton` component based on new `NativeBase` components. Let's take a look at this now:

```
import { Text, Pressable, Center, Icon } from
  "native-base";
import { Ionicons } from "@expo/vector-icons";

export const FooterButton = ({ selected, onPress, title,
  iconName }) => {
  return (
    <Pressable opacity={selected ? 1 : 0.5} py="3" flex={1}
      onPress={onPress}>
      <Center>
        <Icon as={Ionicons} mb="1" name={iconName}
          color="white" size="sm" />
        <Text color="white" fontSize="12">
          {title}
        </Text>
      </Center>
    </Pressable>
  );
};
```

As you can see, we used the `Pressable`, `Center`, and `Icon` components. Using `NativeBase` components is an advantage because it enables you to add styles via props. The `FooterButton` component accepts several props to update icons, active styling, and the label. Here's what the result looks like:

Figure 28.2 – Header and footer

In the next section, you'll learn how to use layout components to organize the content on your screens.

Using layout components

NativeBase provides layout components that simplify the layout code for your screens. You can use these components to build your own grid layouts for the UI components on your screens. Let's take a look at an example. Here are the `App` components:

```
import React from "react";
import { HStack } from "native-base";
import Container from "./Container";
import { CardItem } from "./CardItem";

export default function App() {
  return (
    <Container title="Using Layout Components">
      <HStack space={3} justifyContent="center"
        alignItems="center" mt={2}>
        <CardItem>Card 1</CardItem>
        <CardItem>Card 2</CardItem>
      </HStack>

      <HStack space={3} justifyContent="center"
        alignItems="center" mt={2}>
        <CardItem>Card 3</CardItem>
        <CardItem>Card 4</CardItem>
      </HStack>
    </Container>
  );
}
```

The `HStack` component for horizontal layout and `VStack` for a vertical layout are used to build layouts in `NativeBase` apps. To avoid repeating the code, I defined the `CardItem` component. Now, let's look at it:

```
import { Center } from "native-base";

export const CardItem = ({ children }) => {
  return (
    <Center
      flex={1}
      border="1"
      borderRadius="md"
      bg="gray.200"
      h={50}
      m={1}
      rounded="md"
      shadow={4}
    >
      {children}
    </Center>
  );
};
```

This component has been built on a primitive brick by NativeBase as well. The whole example creates two rows with two columns each; the second row is truncated here because it looks just like the first row. Here's what it looks like:

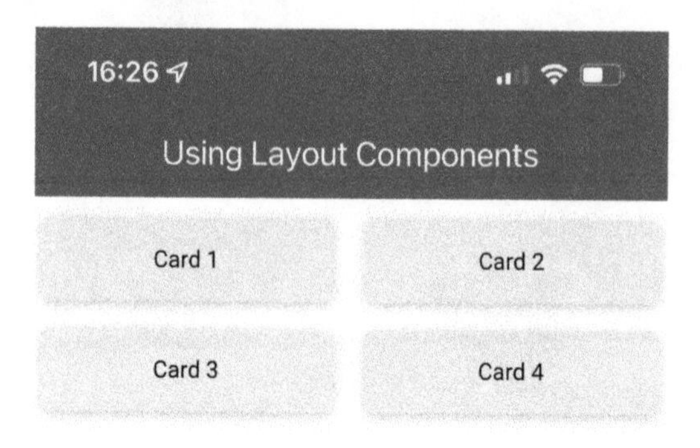

Figure 28.3 – Layout components

Now that you're able to build screen layouts, it's time to use the NativeBase layout components to align form input components.

Collecting input using form components

NativeBase has form components for every type of input imaginable, including the common inputs that you're most likely to use. Form input controls are notoriously difficult for native application developers to use because, even with cross-platform tools such as React Native, the native input controls on the two platforms are so different that you have to write different code for different platforms. With the NativeBase input components, you can usually write your code once. Let's take a look at an example. Here's everything that you need to import:

```
import React, { useState } from "react";
import {
  Input,
  Stack,
  FormControl,
  Select,
  Checkbox,
  Radio,
} from "native-base";
import Container from "./Container";
```

Next, let's look at the state that's used by the various input components to store values collected from the user:

```
const [text, setText] = useState("");
const [select, setSelect] = useState();
const [checkbox, setCheckbox] = useState(true);
const [radio, setRadio] = useState();
const options = ["First", "Second", "Third"];
```

The `text` state defaults to an empty string, the `picker` state defaults to undefined, the `checkbox` state is `false` by default, and the `radio` state defaults to undefined. The `options` array is values used to define the options for the picker input and the radio control. Let's look at how the text input component is used:

```
<Stack>
  <FormControl.Label>Username</FormControl.Label>
  <Input
    p={2}
    placeholder="Username"
```

```
    value={text}
    onChange={setText}
  />
</Stack>
```

The `value` property uses the value state from our `App` component. The `onChange` handler uses `setText()` to update the text state any time the user changes the text input. Next, let's look at the `Select` component:

```
<Stack>
  <FormControl.Label>Select</FormControl.Label>
  <Select
    placeholder="Select"
    selectedValue={select}
    onValueChange={(itemValue) => setSelect(itemValue)}
  >
    {options.map((item) => (
      <Select.Item key={item} label={item} value={item} />
    ))}
  </Select>
</Stack>
```

The `Select` component requires several properties that control the appearance of the dropdown. The `selectedValue` property controls the value of the selector and is set in the picker state. When the picker value changes, the `setPicker()` function updates this state. The options available in the dropdown for the user to choose from are mapped from values in the `options` array to the `Select.Item` components. Let's look at the `Checkbox` component next:

```
<Stack>
  <FormControl.Label>Select</FormControl.Label>
  <Checkbox
    isChecked={checkbox}
    onChange={setCheckbox}
    colorScheme="green"
  >
    Checkbox
  </Checkbox>
</Stack>
```

The `isChecked` visual is controlled by the checkbox state, while the `onChange` handler uses the `setCheckbox()` function to toggle the state of the checkbox. Lastly, let's look at the `Radio` form input control:

```
<Stack>
  <Radio.Group
    name="Radio"
    value={radio}
    onChange={(nextValue) => {
      setRadio(nextValue);
    }}
  >
    <Radio value="one" my={1}>
      One
    </Radio>
    <Radio value="two" my={1}>
      Two
    </Radio>
  </Radio.Group>
</Stack>
```

The `Radio` components are wrapped by `Radio.Group` to control all items. The `value` property controls the selected appearance of the radio and is `true` if the radio state matches the index state of the current radio control. When the user presses one of the radio controls, the `setRadio()` function is called to set the radio state to the index of the radio state that was pressed.

Here's what these controls look like:

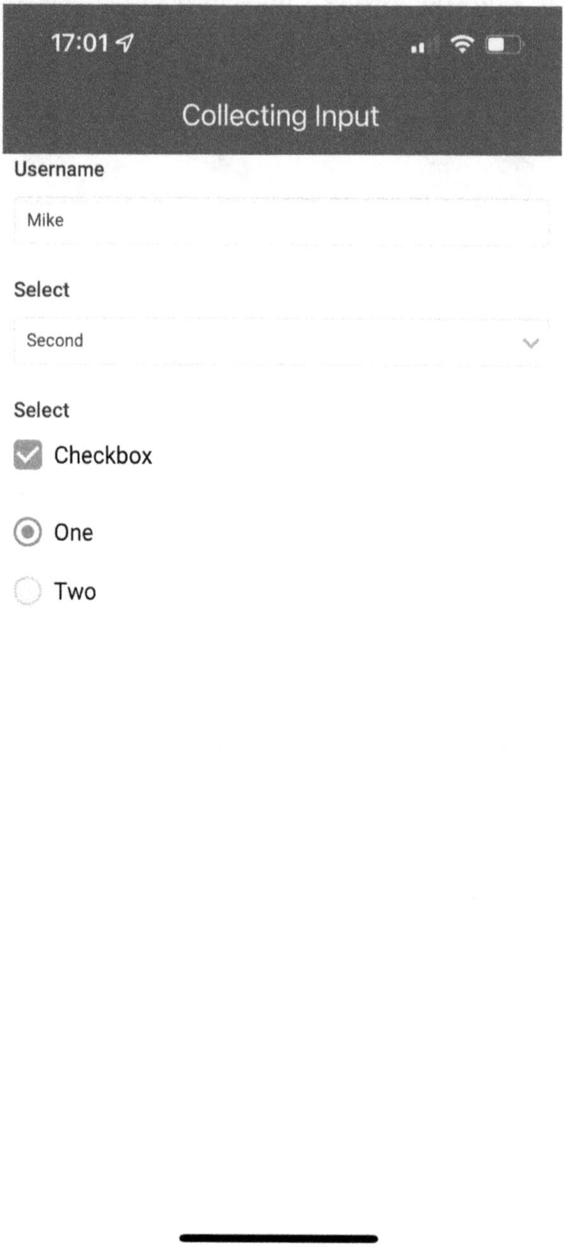

Figure 28.4 – Form components

In this example, we've learned how to use `Input` components to create the form.

Summary

Although we barely scratched the surface of the components available in NativeBase, you now have a sense of what's possible with this library and how it greatly reduces the amount of cross-platform code we need to write. We started by looking at application container components that take care of loading NativeBase fonts and establishing the overall structure for every screen in the app. We then learned about adding navigation and navigation links to our NativeBase app. Then, we organized form components on the screen using NativeBase.

In the next chapter, we'll look at scaling application states in React applications.

Further reading

You can find more information on NativeBase at `https://nativebase.io/`.

Part 3 – React Architecture

In this part, we'll dive into the architecture of a React application. We'll also learn about handling the state of your React application and how to use modern API concepts, such as GraphQL.

This part contains the following chapters:

- *Chapter 29, Handling Application State*
- *Chapter 30, Why GraphQL?*
- *Chapter 31, Building a React GraphQL App*

29
Handling Application State

From early on in this book, you've been using state to control your React components. State is an important concept in any React application because it controls what the user can see and interact with. Without state, you just have a bunch of empty React components.

In this chapter, you'll learn how to handle a more complex application state. Then, you'll learn how to build an architecture that best serves web and mobile architectures. You'll also be introduced to Context, followed by a discussion on the limitations of React architectures, and how you might overcome them using external libraries.

This chapter has the following sections:

- Organizing state in React
- Implementing Context
- State in mobile apps
- Scaling the architecture

Technical requirements

You can find the code files present in this chapter on GitHub at `https://github.com/PacktPublishing/React-and-React-Native-4th-Edition/tree/main/Chapter29`.

Organizing state in React

It can be difficult to think of **user interfaces** (**UIs**) as information architecture. Often, you get a rough idea of how the UI should look and behave and then you implement it. I do this all the time, and it's a great way to get the ball rolling, to discover issues with your approach early, and so on. But then, I like to take a step back and picture what's happening without any widgets. Inevitably, what I've built is flawed in terms of how state flows through the various components. This is fine; at least I have something to work with now. I just have to make sure that I address the information architecture before building too much.

As your application grows in size and there are more pages and components that need to interact with state, the way we have used state in this book might not be sufficient anymore. Therefore, it makes sense to split up your application state into reducers and Context. We can do this with Context, which will be explained in the following sections.

Unidirectionality

So far, we know that state in React is passed down from a parent component to a child component. It flows in one direction, also called **unidirectional**. We've been creating state in UI components and passing it down to other components as props. This is a good approach for a simple state, such as whether a dropdown is opened or a menu should be visible. But, passing props down this way becomes inefficient. As only the direct child component can access the props, there can be situations where you need to pass props through multiple levels of components. This is called **prop drilling**. Prop drilling often occurs when you have lifted state up to a top-level component to make it accessible to more components.

This top-level component has state, but it doesn't render any UI elements. Instead, it renders other React components and passes in their state as properties. Whenever the container state changes, the child components are re-rendered with new property values. This is unidirectional data flow. You can prevent having to lift state and prop drilling with Context.

React takes this idea and applies it to something called Context. Context is an abstract concept that holds an application state. Context lets you pass state down through multiple levels of components. You can combine Context with reducers, as you'll learn later in this chapter. First, I want you to understand why unidirectional data flows are advantageous.

There's a good chance that you've implemented a UI component that changes state, but you're not always sure how it happens. Was it the result of some event in another component? Was it a side-effect from some network call completing? When that happens, you spend lots of time chasing down where the update came from. The effect is often a cascading game of whack-a-mole. When changes can only come from one direction, you can eliminate several other possibilities, thus making the architecture more predictable.

Synchronous update rounds

When you change the state of a React container, it will re-render its children, who re-render their children, and so on. In React, this is called an **update round**. From the time state changes to the time that the UI elements reflect this change, this is the boundary of the round. It's nice to be able to group the dynamic parts of application behavior into larger chunks to have a clearer overview.

A potential problem with state in React components is that they can interweave with one another and render in a non-deterministic order. For example, what if some API call completes and causes a state update to happen before the rendering has completed in another update round? The side effects of asynchronicity can accumulate and morph into unsustainable architectures if not taken seriously.

With Context, you get more control over update rounds. JavaScript is a single-threaded, run-to-completion environment that should be embraced by working with it rather than against it. Update the whole UI, and then update the whole UI again. It turns out that React is a really good tool for this job.

Predictable state transformations

When using Context, you have one or multiple Context instances that hold the application state. A change of the values in the Context happens synchronously and unidirectionally, making the system as a whole more predictable and easier to reason about. However, there's still one more thing you can do to ensure that side effects aren't introduced.

You're keeping all your application state in Context instances, which is great, but you can still break things by mutating data in other places. These mutations might seem innocent at first glance, but they're toxic to your architecture. For example, the callback function that handles a `fetch()` call might manipulate the data before passing it to the Context value. Or, an event handler might generate some structure and pass it to the Context. There are limitless possibilities. For this reason, you should combine Context with reducers from the `useReducer()` Hook of React.

The problem with performing these state transformations outside a Context instance is that you don't necessarily know that they're happening. Think of mutating data as a butterfly effect: one small change has far-reaching consequences, which aren't obvious at first. The solution is to only mutate state in the Context, without exception. It's predictable and easy to trace cause and effect of your React architecture this way.

Unified information architecture

Let's take a moment to recap the ingredients of our application architecture so far:

- **React Web**: Applications that run in web browsers
- **React Native**: Applications that run natively on mobile platforms
- **Context**: Way of handling the scalable state in React applications

Remember, React is just an abstraction that sits on top of a render target. The two main render targets are browsers and native mobile. This list will likely grow, so it's up to you to design your architecture in a way that doesn't exclude future possibilities. The challenge is that you're not porting a web application to a native mobile application; they're different applications, but they serve the same purpose.

Having said that, is there a way that you can still have a unified information architecture that can be used by these different applications? The best answer I can come up with, unfortunately, is sort of. You don't want to let the different web and mobile user experiences lead to drastically different approaches in handling state. If the goals of the applications are the same, then there must be some common information that you can share, using the same Context.

The difficult part is the fact that web and native mobile are different experiences, which means that the shape of your application state will be different. It must be different; otherwise, you would just be porting from one platform to the other, which defeats the purpose of using React Native to leverage capabilities that don't exist in browsers.

Now you know everything about state management in React and Context, let's continue by implementing Context in an application in the next section.

Implementing Context

With Context, we can implement the application state, as you'll learn in this chapter. To implement Context, we'll create a basic React application that needs state. The application itself will be a news application. It's a simple app, but I want to highlight the architectural challenges as I walk through the implementation. Even simple apps get complex when you're paying attention to what's going on with the data.

We'll build the web version, and then you could implement the same patterns on a React Native app for iOS and Android. You'll see how you can share architectural concepts between your apps. This lowers the conceptual overhead when you need to implement the same application on several platforms.

You're implementing two apps right now, but this will likely be more in the future as React expands its rendering capabilities.

Creating Context

The basis of implementing Context for a React application is importing the `createContext()` Hook. With this Hook, you can create a Context instance, which is needed to store the state information in your application. You create a Context for every main slice of your application. So, for example, your app would have an `App` Context and an `Articles` Context instance:

```
import { createContext } from 'react';

export const ArticlesContext = createContext();

const initialState = {
  article: {},
  articles: [],
  filter: '',
  loading: true,
  error: '',
};
```

The Context has an initial state, which is useful when the application first starts. This is enough to render components, but that's about it. Once the user starts interacting with the UI, you need a way to change the state of the store. Context works best when combined with a reducer from the `useReducer()` Hook in React. You assign a reducer function to each slice of state in your store. This is shown later, in the *Reducer functions* section.

Context provider

Context has a `Provider` component, which is used to wrap the top-level components of your application. This will ensure that state data is available to every component in your application.

In the hipster newsreader app you're developing, you'll wrap the `Router` component from React Router with a `Provider` component. Then, as you build your components, you know that the data from the Context will be available. The `App` component includes the page heading and a list of links to various article categories, which serves as a filter. When the user moves around the UI, the `App` component is always rendered, but each `Route` element renders different content based on the current route. Here's what the `App` component looks like:

```
import { Routes, Route, BrowserRouter } from 'react-router-
dom';
import Filter from './components/Filter';
import Home from './components/Home';
import Article from './components/Article';
import AppContext from './context/AppContext';

function App() {
  return (
    <BrowserRouter>
      <div>
        <h1>Hipster news app</h1>
        <AppContext>
          <Filter />
          <Routes>
            <Route path='/' element={<Home />} />
            <Route exact path='/articles/:id'
              element={<Article />} />
          </Routes>
        </AppContext>
      </div>
    </BrowserRouter>
  );
}

export default App;
```

In the App component, AppContext is imported. This is a wrapper that wraps all the Provider components of all the Context instances of the application. Wrapping the Provider components in one component makes it easier to add new Context instances to your application at a later stage. The AppContext components look like the following:

```
import { ArticlesContextProvider } from
  './ArticlesContext';

const AppContext = ({ children }) => {
  return <ArticlesContextProvider>{children}
    </ArticlesContextProvider>;
};

export default AppContext;
```

Every component within the component tree that is wrapped with AppContext can now access the state that is available from Context. This means that when the Context changes, the state data is automatically passed to each application component.

Reducer functions

To make Context updates more stable, you can combine them with reducer functions. Reducer functions look for state updates and return a new value for the state or content. Every Context instance created with the createContext() Hook has its own reducer functions.

Reducer functions take the current state and an action that is passed to it. Based on this action, it will change the state value. The reducer is often a switch statement that iterates over a map of actions, which are defined in uppercase. The reducer for ArticlesContext has the following format:

```
import { createContext, useReducer } from 'react';

export const ArticlesContext = createContext();

const initialState = {
  article: {},
  articles: [],
  filter: '',
```

```
    loading: true,
    error: '',
  };

  const reducer = (state, action) => {
    switch (action.type) {
      case 'GET_ARTICLES_SUCCESS':
        return {
          ...state,
          articles: action.payload,
          loading: false,
        };
      case 'GET_ARTICLES_ERROR':
        return {
          ...state,
          articles: [],
          loading: false,
          error: action.payload,
        };
      default:
        return state;
    }
  };
```

Reducer functions are passed to a `useReducer()` Hook together with the initial state. The output of this Hook will be the current state value after executing the reducer functions, and a `dispatch()` function. Together with the Context instance, the `Provider` component that is available on this Context instance is important. With this `Provider` component, you set the value for the Context, which is the state value:

```
  export const ArticlesContextProvider = ({ children }) => {
    const [state, dispatch] = useReducer(reducer,
      initialState);

    async function fetchArticles(filter = '') {
```

```
    try {
      const data = await fetch(
        'http://localhost:3001/articles${filter ?
          '?category=${filter}' : ''}',
      );
      const result = await data.json();

      if (result) {
        dispatch({ type: 'GET_ARTICLES_SUCCESS',
          payload: result });
      }
    } catch (e) {
      dispatch({ type: 'GET_ARTICLES_ERROR',
        payload: e.message });
    }
  }

  return (
    <ArticlesContext.Provider
      value={{ ...state, fetchArticles }}
    >
      {children}
    </ArticlesContext.Provider>
  );
};

export default ArticlesContext;
```

The `dispatch()` function is how these action creator functions can deliver payloads to the Context. For example, the `fetchArticles()` function is a call directly to `dispatch()` and is called when the user wants to load the list of articles. However, the `fetchArticles()` call involves asynchronous behavior. This means that `dispatch()` isn't called until the `fetch()` promise resolves. It's up to you to make sure that nothing unexpected happens in between. We'll look at the `Home` component next to see how this works.

The Home component

The `Home` component is available on `http://localhost:3000` and shows the list of all available articles. It also has a filter of the different categories of articles that can be selected, but this filter is rendered by the `App` component. Let's break down how the `Home` component is structured:

```
import React, { useContext, useEffect } from 'react';
import { Link } from 'react-router-dom';
import ArticlesContext from '../context/ArticlesContext';

function Home() {
  const { fetchArticles, articles } =
    useContext(ArticlesContext);

  useEffect(() => {
    if (!articles.length) {
      fetchArticles();
    }
  }, [articles, fetchArticles]);

  return (
    <ul style={listStyle}>
      {articles.length === 0 ? <li
        style={listItemStyle}>...</li> : null}
      {articles.map(({ id, title, summary }) => (
        <li key={id} style={listItemStyle}>
          <Link to={'articles/${id}'}>
            <button style={titleStyle}>{title}</button>
          </Link>

          <p>
            <small>
              <span>{summary} </span>
              <Link to={'/articles/${id}'}>More...</Link>
```

```
                </small>
              </p>
            </li>
          ))}
        </ul>
      );
    }

    export default Home;
```

This component is using the `useContext()` Hook. With this Hook, you can read the Context value from a Context instance, such as `ArticlesContext`. The Context value has the articles that we want to render in the UI and the `fetchArticles()` function. This function gets the articles from an API from within a `useEffect()` Hook, when there aren't any articles in the Context yet.

Preventing unwanted re-renders

When the value for the Context changes, all components within the component tree might re-render. If multiple components are wrapped within a `Provider` component, these components are looking for state updates within that Context.

For this application, this means that every component nested in `AppContext` might re-render if one of the Context instances is updated. This is something you want to prevent, as components should only re-render when the values they consume are changed. In the following code block, you can see the implementation of this:

```
export const ArticlesContextProvider = ({ children }) => {
  const [state, dispatch] = useReducer(reducer,
    initialState);

  async function fetchArticles(filter = '') {
    try {
      const data = await fetch(
        'http://localhost:3001/articles${filter ?
          '?category=${filter}' : ''}',
      );
```

```
      const result = await data.json();

      if (result) {
        dispatch({ type: 'GET_ARTICLES_SUCCESS',
          payload: result });
      }
    } catch (e) {
      dispatch({ type: 'GET_ARTICLES_ERROR',
        payload: e.message });
    }
  }

  const value = useMemo(
    () => ({ ...state, fetchArticles }),
    [state],
  );

  return (
    <ArticlesContext.Provider value={value}>
      {children}
    </ArticlesContext.Provider>
  );
};

export default ArticlesContext;
```

In this code block, the value for Context is wrapped in a `useMemo()` Hook, so you prevent unwanted re-renders for your application.

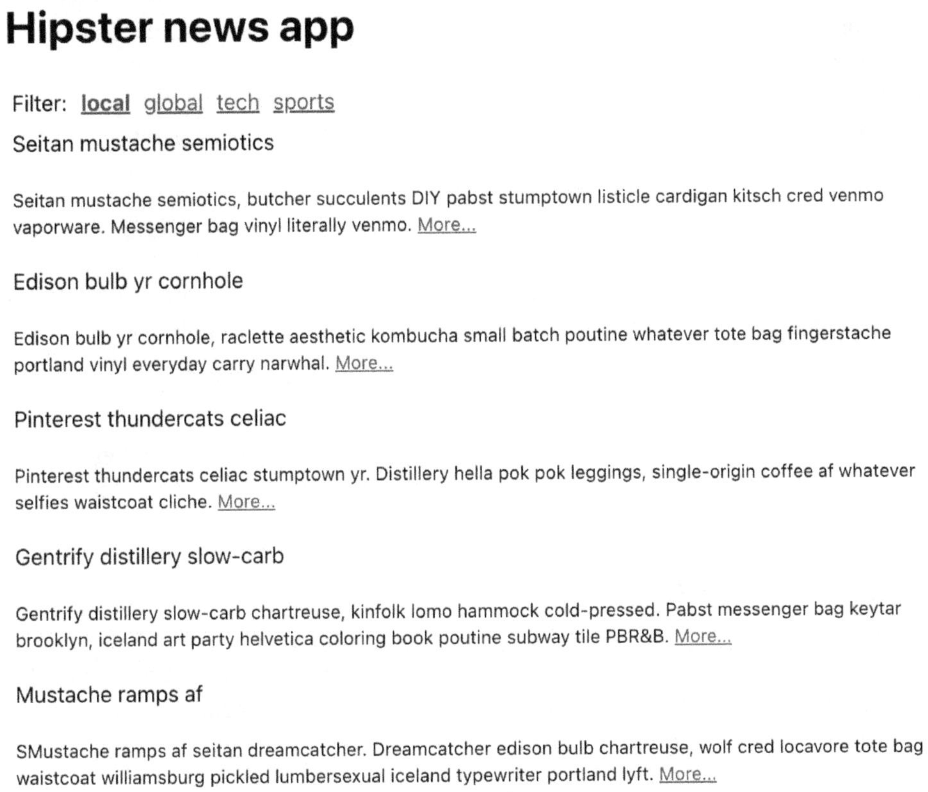

Hipster news app

Filter: **local** global tech sports

Seitan mustache semiotics

Seitan mustache semiotics, butcher succulents DIY pabst stumptown listicle cardigan kitsch cred venmo vaporware. Messenger bag vinyl literally venmo. More...

Edison bulb yr cornhole

Edison bulb yr cornhole, raclette aesthetic kombucha small batch poutine whatever tote bag fingerstache portland vinyl everyday carry narwhal. More...

Pinterest thundercats celiac

Pinterest thundercats celiac stumptown yr. Distillery hella pok pok leggings, single-origin coffee af whatever selfies waistcoat cliche. More...

Gentrify distillery slow-carb

Gentrify distillery slow-carb chartreuse, kinfolk lomo hammock cold-pressed. Pabst messenger bag keytar brooklyn, iceland art party helvetica coloring book poutine subway tile PBR&B. More...

Mustache ramps af

SMustache ramps af seitan dreamcatcher. Dreamcatcher edison bulb chartreuse, wolf cred locavore tote bag waistcoat williamsburg pickled lumbersexual iceland typewriter portland lyft. More...

Figure 29.1 – Our news app with mock data

All the data that is pulled from the API will be managed in the Context. The news application you've built in this section can also be transformed to work as a mobile app, as you'll learn in the next section.

Managing state in mobile apps

What about using Context in React Native mobile apps? Of course, you should if you're developing the same application for the web and for native platforms. In fact, you could copy and paste all the Context-related code to a React Native application and use it to handle state. I encourage you to download the code for this book and convert the application from this chapter to React Native as an exercise.

There really is no difference in how you use Context in a mobile app. The only difference is in the shape of the state that's used. In other words, don't think that you can use the exact same Context and reducer functions in the web and native versions of your app. Think about React Native components. There's no one-size-fits-all component for many things. You have some components that are optimized for the iOS platform, while others are optimized for the Android platform. It's the same idea with Context. The information that you want the mobile UI to render might be different from what you want to render on the web.

For mobile apps, you also must deal with an offline state, as users could also open the mobile application without an internet connection. Persisting state is, therefore, more important on mobile compared to the web. To persist state, you can use external React libraries.

You can use what you've learned about Context in this chapter with your React Native application too. I'd recommend that you make a React Native version of the news app from this chapter as a challenge. The next section discusses how to scale these applications.

Scaling the architecture

By now, you probably have a pretty good grip of using Context, combining it with reducer functions, and using it to implement sound information architecture for React applications. The question then becomes, how sustainable is this approach, and can it handle arbitrarily large and complex applications?

Context is a good way to handle the state for your application if this state isn't continuously updated. By dividing your Context into different smaller Context instances, it becomes more scalable. You can predict what's going to happen as the result of any given action because everything is explicit. It's declarative, it's unidirectional, and without side effects. But, it isn't without challenges.

The limiting factor with Context is also its bread and butter; because everything is explicit, applications that need to scale up in terms of feature count and complexity ultimately end up with more moving parts. There's nothing wrong with this; it's just the nature of the game. The unavoidable consequence of scaling up is slowing down. You simply cannot grasp enough of the big picture in order to implement things quickly.

Context is a pure client-side approach to handling state. If you want to handle state from the server side or handle state that only comes from external data sources, have a look at React Query. It's an open source library to handle state, caching, data persistency, and much more.

In the final two chapters of this book, we're going to look at a related but different approach to Context: GraphQL. I believe that this technology can scale in ways that Context cannot, as it resides directly at the level of your data.

Summary

In this chapter, you learned about Context. With Context, you can handle state for any React application. In combination with reducer functions, you can use it for building information architecture for your React application. State in React involves unidirectional data flow, synchronous update rounds, and predictable state transformations.

Next, we walked through a detailed implementation of a React application that uses Context. Context provides an implementation for complex state situations, the benefit of which is predictability everywhere.

Then, you learned whether Context has what it takes to build scalable architectures for our React applications. The answer is yes, for the most part. For the remainder of this book, however, we're going to explore GraphQL to see whether these technologies can scale your applications to the next level. In the next chapter, you'll learn about the concepts of GraphQL and what it means for a React application.

Further reading

For more information, check out the following links:

- Context: `https://reactjs.org/docs/context.html`
- React Query: `https://react-query.tanstack.com/`

30
Why GraphQL?

In the preceding chapter, you learned about the architectural principles of state management in React. In particular, you used Context to implement more complex state management in a React application. Context, in combination with reducer functions, will help you to understand how state changes and flows in your application are a good thing. At the end of the preceding chapter, you also learned about the potential limitations in terms of scale.

In this chapter, we are going to walk you through yet another approach to handling state in a React application. Similar to Context, GraphQL can be used with both web and mobile React applications. GraphQL is a query language for APIs and is implemented on the server side. To connect with GraphQL from a React application, we'll be using the library called Apollo Client.

Unlike Context, you don't have to write reducers and actions to deal with state management. Instead, GraphQL provides a more declarative way of handling data fetching and the state of that data in your application afterward. For this, GraphQL can be used with Apollo Client, which provides you with components and hooks to fetch and mutate data from any GraphQL server.

In the final chapter of this book, you'll work on a React implementation of the ever-popular TodoMVC application using GraphQL with Apollo Client.

In this chapter, you'll learn about the following:

- The need for another approach to handle data in React apps
- The high-level vocabulary of GraphQL
- Declarative data fetching
- Mutations to update data

Approaching state with GraphQL

How to handle application state with GraphQL? This was the exact question I had when I learned of GraphQL. Then, I reminded myself that the beauty of React is that it's just the view abstraction of the UI. Of course, there are going to be many approaches to handling data. So, the real question is, what makes using GraphQL better or worse than using something such as Context for state management and data fetching?

At a high level, how GraphQL is handled with Apollo Client is similar to what we've done previously in this book. At a more practical level, the value of Apollo Client is its ease of implementation. For example, with Context, you have a lot of implementation work to do just to populate the stores with data. This gets verbose over time, as it's difficult to scale Context beyond a certain point if you've got to write that much code for every new feature that you want to implement.

It's not the individual data points that are difficult to scale. Rather, it's the aggregate effect of having lots of fetch requests that end up building very complicated stores. Apollo Client changes this by allowing you to declare the data that a given component needs and letting Apollo Client figure out the best way to fetch this data and synchronize it with the local store. Under the hood, it will use a similar logic to what you've already written in the previous chapters.

Is the Apollo and GraphQL approach better than Context and other approaches for handling data in React applications? In some respects, yes, it is. Is it perfect? Far from it. There is a learning curve involved and not everyone is able to deal with it. It's immutable and parts of it are difficult to use. However, just knowing the premise of how GraphQL works with Apollo Client and seeing it in action is worth your while, even if you decide against it.

Now, let's pick apart some vocabulary.

Understanding some verbose vernacular about GraphQL

Before I start going into more depth about data dependencies and mutations, I think it makes sense for me to throw some general Relay and GraphQL terminology and definitions out there:

- **GraphQL**: This is a query language that is used to specify data requirements and data mutations.
- **Apollo Client**: This is a library that manages application data fetching and data mutations. It provides higher-order components and hooks that feed data to our application components. Also, it comes with React Hooks support and caching out of the box.
- **Query**: This is a part of a data dependency, expressed in the GraphQL syntax, and executed by an encapsulated Relay mechanism.
- **Fragment**: This is a part of a larger GraphQL query.
- **Mutation**: This is a special type of GraphQL query that changes the state of a remote resource. Apollo Client has to figure out how to reflect this change in the frontend once it completes.
- **Subscription**: This is a GraphQL type used for real-time events between the server and the client application; for example, for notifications or chat messages.

Let's quickly talk about declarative data fetching and mutations before we move on to look at some application code.

Declarative data fetching

As mentioned earlier, GraphQL is a query language that lets you define what the response of an API looks like by how you structure your query. This is a more declarative approach to data fetching than you see in other APIs. Not only is it a query language, but it also provides a runtime to fulfill those queries based on your existing data. Also, not only can you use GraphQL to fetch data with queries, but you can also send mutate data by using mutations.

When you want to use GraphQL, the API that you're using for data fetching should support GraphQL. This means the server should have a schema that describes which operations (that is, queries, mutations, or subscriptions) are allowed and which data fields can be requested. Every operation that is described in the schema for a GraphQL server can be executed by sending a document containing these operations. Other than with REST APIs, you have complete control over the shape of your data as you define what structure the response should have in your operation.

Let's get a taste of how GraphQL queries work. If you want to display the first and last name of a user, you need to tell the GraphQL server that you want to retrieve these fields. Then, you can be sure that the data will always be there. Here's an example of what a query looks like:

```
query getUser {
  user {
    firstName
    lastName
  }
}
```

In this query, you have described that you want to retrieve the `firstName` and `lastName` fields for a user. When you send this query in a document to a GraphQL server, it will respond with a JSON object containing these fields (and only these fields):

```
"data": {
  "user": {
    "firstName": "John",
    "lastName": "Doe"
  }
}
```

The preceding request is similar to how a REST API would handle a request, for example, a call to a `/users` endpoint. What differs is the shape of the data that is returned by the GraphQL server and the fact that you can use Apollo Client to retrieve this data.

Apollo Client can be used for data fetching in different ways, using React concepts that you've already explored in this book. One of those ways is by using hooks to execute GraphQL operations, such as sending a query. For this, you can use the `useQuery()` Hook from Apollo Client, which not only sends the query but also handles state management for you.

Let's see how to use a `useQuery()` Hook to retrieve data:

```
import { gql, useQuery } from '@apollo/client';

const GET_USER = gql'
  query getUser {
    user {
      firstName
```

```
      lastName
    }
  }
';

function User() {
  const { loading, error, data } = useQuery(GET_USER);

  if (loading) return 'Loading...';
  if (error) return 'Error! ${error.message}';
  const { firstName, lastName } = data.user;

  return (
    <p>
      'Hi there, ${firstName} ${lastName}'
    </p>
  );
}

export default User;
```

The `useQuery()` Hook takes a GraphQL query as a prop and returns an object with the `loading`, `error`, and `data` state variables. When there is no data fetched yet, the `loading` variable will be true. As soon as the data is loaded, the `error` or `data` variables will be resolved with information from the GraphQL server.

Depending on the schema of the GraphQL server, you can add more fields to the query or even query nested relationships for this user. If the GraphQL schema allows for nested relationships, you can define these in your query as follows:

```
query getUser {
  user {
    firstName
    lastName
    todos {
      title
      status
```

```
      }
    }
  }
```

The preceding query will also retrieve the `todos` field for this user, along with `firstName` and `lastName`.

This query will return the following data:

```
  "data": {
    "user": {
      "firstName": "John",
      "lastName": "Doe",
      "todos": [
        {
          "title": "Do dishes",
          "status" "complete"
        },
        {
          "title": "Walk the dog",
          "status": "open"
        }
      }
    }
  }
```

Here, you can see how the `todos` object is not only shaped exactly as it was defined in the query, but also as a list. If you have a REST API, you will need to send two different requests to two different endpoints to retrieve this. For example, you can send one request to the `/users` endpoint and another request to the endpoint that returns the list of todos for a user.

Don't dwell on the Apollo Client and GraphQL specifics just yet. The idea here is to simply illustrate that this is what you need to write to get data from a GraphQL server. The rest is just bootstrapping Apollo Client for data fetching and state management, which you'll see in the next chapter.

Mutating application state

GraphQL mutations are the actions that cause side effects in your systems because they change the state of a particular resource that your UI cares about. What's interesting about mutations is that they care about side effects that happen to your data as a result of a change in the state of something. For example, if you change the information of a user, this will certainly impact the screen that displays the user information. However, it could also impact a listing screen that shows the information of several users.

Let's see what a mutation looks like:

```
mutation changeTodoStatus($input: ChangeTodoStatusInput!) {
  changeTodoStatus(input: $input) {
    todo {
      title
      status
    }
    user {
      todos {
        title
        status
      }
    }
  }
}
```

This mutation will change the status of a `todo` item and return the updated information of that item. But that's not all this mutation does; it also returns the user information containing all the todos of that user. When the status of a `todo` item changes, a screen that shows the todos for this user might also change. This is how Apollo Client and GraphQL can determine what might be affected as a side effect of performing this mutation, as the updated information for the user will also be returned.

Similar to how you used Apollo Client to retrieve user information, this information can also be mutated using Apollo Client. Again, there are multiple approaches to using GraphQL mutations with Apollo Client. The default way to handle mutations is by using the `useMutation()` Hook:

```
import { gql, useMutation} from '@apollo/client';

const ADD_TODO = gql'
  // Insert mutation
';

function AddTodo() {
  let input;
  const [addTodo, { data, loading, error }] =
    useMutation(ADD_TODO);

  if (loading) return 'Submitting...';
  if (error) return 'Submission error! ${error.message}';

  return (
    <div>
      <form
        onSubmit={e => {
          e.preventDefault();
          addTodo({ variables: { text: input.value } });
          input.value = '';
        }}
      >
        <input
          ref={node => {
            input = node;
          }}
        />
        <button type="submit">Add Todo</button>
      </form>
    </div>
  );
```

```
}

export default AddTodo;
```

The `useMutation()` Hook takes the mutation as a parameter and returns an object to execute the mutation and an object with the data that will be returned. With a simple form element, the mutation function can be called when this form gets submitted. When it returns the data for the user, a *Completed!* message will be shown on the screen.

You'll see more mutations in action in the following chapter, where you'll implement Apollo Client in a React application.

Summary

The goal of this chapter was to briefly introduce you to the concepts of GraphQL and Apollo Client prior to the final chapter of this book, where you're going to implement some Apollo Client and GraphQL code.

Apollo Client is yet another approach to the state management problem in React applications. It's different in the sense that it reduces the complexities associated with the data fetching code that we must write with other approaches such as Context.

The two key aspects of Apollo Client are declarative data fetching and explicit mutation side-effect handling. All of this is expressed through the GraphQL syntax and React Hooks. In order to have an Apollo Client application, you need a GraphQL backend from which the data can be retrieved.

Now, let's move on to the final chapter, where you'll examine GraphQL concepts in more detail by creating a React application with Apollo Client.

Further reading

Refer to the following links for more information:

- GraphQL: `https://graphql.org/`
- Apollo Client: `https://www.apollographql.com/docs/react/`

31
Building a GraphQL React App

In the previous chapter, you received an extensive introduction to Apollo and GraphQL and learned why and how you should use this approach for your React application.

Now, you can build your Todo React application using Apollo Client. By the end of this chapter, you should be comfortable with knowing how data moves around in a GraphQL-centric application.

In this chapter, we'll cover the following topics:

- Todo and Apollo Client
- The GraphQL schema
- Bootstrapping Apollo Client
- Adding to-do items
- Rendering to-do items
- Completing to-do items

Technical requirements

You can find the code files present in this chapter on GitHub at `https://github.com/PacktPublishing/React-and-React-Native-4th-Edition/tree/main/Chapter31`.

Creating a Todo app

In this chapter, we'll build a Todo example for React that uses GraphQL to handle its data. This example is based on a popular open source library (`https://todomvc.com/examples/react/`), which is a robust, yet concise, starting point for creating the Todo application for this chapter.

I'm going to walk you through an example React implementation of a Todo app. Also, you can find a React Native implementation of this same web app in the GitHub repository for this chapter. The key is that the mobile version will use the same GraphQL backend as the web UI. I think this is a win for React developers who want to build both web and native versions of their apps as they can share the same schema!

The code for this chapter contains both a web version build with React and a native version with React Native. If you've worked on frontend development in the past 5 years, you've probably come across a sample Todo app. Here's what the web version looks like:

Figure 31.1 – A Todo MVC app example

Implementing GraphQL for this Todo app works similarly for both the web and mobile versions. We won't focus on both, only the web version. Apollo Client works mostly the same on native platforms as it does on web platforms, and the GraphQL backend can be shared between web and native apps. The next section shows what this GraphQL backend looks like.

Constructing a GraphQL schema

The schema is the vocabulary used by the GraphQL backend server and the Apollo Client implementation in the frontend. The GraphQL type system enables the schema to describe the data that's available and how to put it all together when a query request comes in. This is what makes the whole approach so scalable – the fact that the GraphQL runtime figures out how to put data together. All you need to supply are functions that tell GraphQL where the data is – for example, in a database or a remote service endpoint.

Let's take a look at some of the types used in the GraphQL schema for the Todo app. We'll start with `Todo` itself:

```
type Todo {
  id: ID!
  text: String!
  complete: Boolean
}
```

This type describes the `Todo` objects used throughout the application, including all the optional and required fields for this type. In the example code, you can see that the types followed with an exclamation mark are required (`id` and `text`), and the ones without an exclamation mark are optional (`complete`). Everything else in the GraphQL schema is based on this type, either directly or indirectly.

Next, let's look at the types that tie the `Todo` type to the user who is interacting with the app:

```
type User {
  id: ID!
  totalCount: Int!
  completedCount: Int!
  todos: [Todo]!
}
```

By implementing these types, the end result is a `User` type with a list of todos, including the total number of todos and the total number of completed todos. This type can be accessed from our components, which we'll see shortly. Our schema also needs to declare how data changes.

Let's look at some of the mutation types used with this app:

```
type Mutation {
  addTodo(text: String): [Todo]
  changeTodoStatus(id: Int!, complete: Boolean): [Todo]
  markAllTodos: [Todo]
  removeCompletedTodos: [Todo] removeTodo(id: Int!): [Todo]
  renameTodo(id: Int!, text: String): [Todo]
}
```

Each mutation type describes what it takes as input and what the resulting payload looks like once the operation is complete. The mutation type ties all of this together and provides an interface for everything that changes in the application. In this schema, there are mutations to add new todos, change the name or status of a todo, or delete todos. These mutations will be used when creating the application later on.

Now that there's a GraphQL schema in place, we're ready to put it into our application using Apollo and React in the next section.

Bootstrapping the Apollo Client

At this point, you will have the GraphQL backend up and running. Now, you can focus on your React components in the frontend. In particular, you're going to look at the Apollo Client in a React application. In web apps, it's usually React Router that bootstraps Apollo Client. Let's now look at the `src/index.js` file that serves as the entry point for your web app:

```
import React from 'react';
import ReactDOM from 'react-dom/client';
import { ApolloClient, InMemoryCache, ApolloProvider } from
  '@apollo/client';
```

```
import App from './App';
import './index.css';

const client = new ApolloClient({
  cache: new InMemoryCache(),
  uri: 'http://localhost:4000/graphql',
});

const root =
  ReactDOM.createRoot(document.getElementById('root'));
root.render(
  <React.StrictMode>
    <ApolloProvider client={client}>
      <App />
    </ApolloProvider>
  </React.StrictMode>,
);
```

The `ApolloProvider` component is wrapping the `App` component in the same way as with Context. By wrapping a component tree with a `Provider`, every component in the tree can connect with the GraphQL server. The GraphQL server is running at `http://localhost:4000/graphql`, and you can use this address to set up an Apollo Client instance for the application.

Also, `InMemoryCache` from `@apollo/client` is used to get caching for your application data out of the box:

```
const client = new ApolloClient({
  cache: new InMemoryCache(),
  uri: 'http://localhost:4000/graphql',
});
```

This is how you communicate with the GraphQL backend – by configuring a client. However, in the React Native example, you're using the network IP address of your computer, as Expo doesn't accept a localhost address. Using your local network IP address means all requests to the GraphQL backend are being made on your machine. To get your network IP address (or IPv4 address) of your local machine to access the GraphQL backend, follow these steps:

- *For Windows*: Open the terminal (or Command Prompt) and run this command:

```
ipconfig
```

  This will return a list, as follows, with data from your local machine. In this list, you need to look for the **IPv4 Address** field:

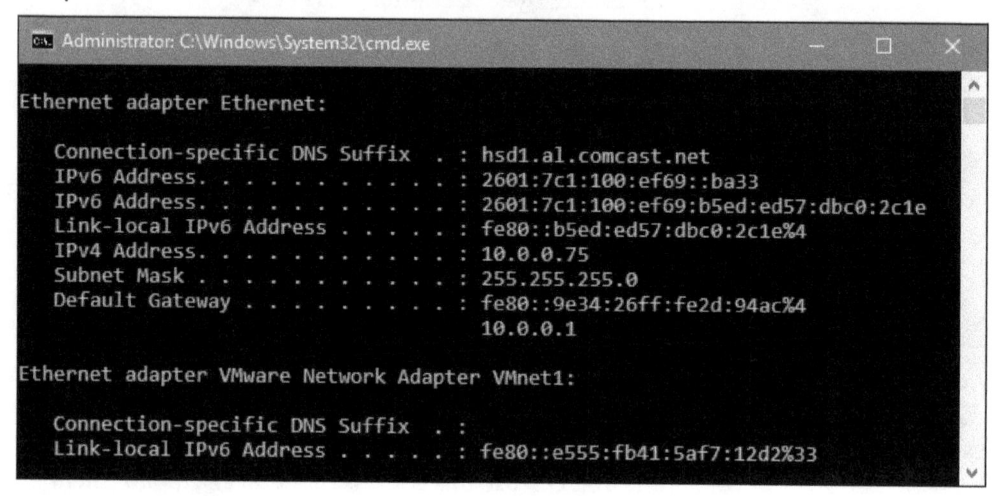

Figure 31.2 – Getting the network address on Windows

- *For macOS*: Open the Terminal and run this command:

```
ipconfig getifaddr en0
```

  After running this command, the IPv4 address of your machine gets returned, which looks like this:

```
192.168.1.107
```

  You can use this address to set up an Apollo Client instance for the React Native application if you decide to work on the React Native example.

Let's get back to implementing Apollo Client in the React web application. The next step is to implement a `useQuery()` Hook. This Apollo Client Hook is used to get data from the server using GraphQL queries. It expects a query property, which is used to get the todo items for a user in `src/App.js`:

```
import { useQuery } from '@apollo/client';

import TodoList from './components/TodoList';
import TodoListFooter from './components/TodoListFooter';
import { GET_USER } from './constants';

function App() {
  const { loading, error, data } = useQuery(GET_USER, {
    variables: {
      userId: 'me', // Mock authenticated ID that matches
        database
    },
  });

  if (loading) return 'Loading...';
  if (error) return 'Error! ${error.message}';

  const hasTodos = data.user.totalCount > 0;

  return (
    <div>
      <section className='todoapp'>
        <header className='header'>
          <h1>todos</h1>
        </header>

        <TodoList user={data.user} />
        {hasTodos && <TodoListFooter user={data.user} />}
      </section>
    </div>
  );
}

export default App;
```

In this code block, the GraphQL backend returns a set of todo items by calling a query with the `useQuery` Hook. The value for the query can be found in the `src/constants.js` file, which hosts all the queries and mutations for this application:

```
import { gql } from '@apollo/client';

export const GET_USER = gql'
  query GetUser($userId: String) {
    user(id: $userId) {
      id
      totalCount
      completedCount
      todos {
        id
        text
        complete
      }
    }
  }
';
```

As you can see, this query requires the `userId` parameter. This value is passed from the second parameter in the `useQuery()` Hook.

Then, the Hook will return a `loading` value, once the query is transferred to the GraphQL backend, and the `error` and `data` values when the GraphQL data is ready:

```
const { loading, error, data } = useQuery(GET_USER, {
  variables: {
    userId: 'me', // Mock authenticated ID that matches
      database
  },
});
```

If something went wrong, `error` will contain information about the error, and you can return a message to the user. Otherwise, you can return the components that need the data value. If there's no error and no props, it's safe to assume that the GraphQL data is still loading.

Next, we'll have a look at using mutations to add new todo items to the application.

Adding todo items

In the `App` component, there's also a text input that allows the user to enter new todo items. When they're done entering the todo item, Apollo Client will need to send a mutation to the backend GraphQL server. Here's what the component code looks like:

```
function App() {
  const [addTodo] = useMutation(ADD_TODO, {
    refetchQueries: [
      {
        query: GET_USER,
        variables: {
          userId: 'me',
        },
      },
    ],
  });

  return (
    <div>
      <section className='todoapp'>
        <header className='header'>
          <h1>todos</h1>

          <TodoTextInput
            className='new-todo'
            onSave={(text) => addTodo({ variables: { text }
              })}
            placeholder='What needs to be done?'
          />
        </header>

        <TodoList user={data.user} />
        {hasTodos && <TodoListFooter user={data.user} />}
      </section>
```

```
    </div>
  );
}

export default App;
```

It doesn't look that different from your typical React component. The method to add a new todo item is created by the `useMutation()` Hook. This Hook uses a mutation, which is how you tell the GraphQL backend that you want to create a new todo item. This Hook looks very similar to the `useMutation()` Hook that you saw in the previous section. This Hook needs a mutation, which is the `ADD_TODO` mutation that you can find in the `src/constants.js` file:

```
import { gql } from '@apollo/client';

export const ADD_TODO = gql'
  mutation AddTodo($text: String) {
    addTodo(text: $text) {
      id
    }
  }
';
```

This mutation takes just one variable, which you can pass to the mutation by using the `addTodo` callback function that was returned by the `useMutation()` Hook. You can call this function when the user submits something in the input field in the `TodoInput` component:

```
<TodoTextInput
  className='new-todo'
  onSave={(text) => addTodo({ variables: { text } })}
  placeholder='What needs to be done?'
/>
```

When the mutation has been sent to the GraphQL backend, this same Hook can be used to refetch any queries that are defined in your application. If a new todo is added using the mutation, you want your user to see the new list of todos by refetching the `GET_USER` query from the `App` component. To refetch a query, you can pass a value for `refetchQueries` to the `useMutation()` Hook:

```
const [addTodo] = useMutation(ADD_TODO, {
  refetchQueries: [
    {
      query: GET_USER,
      variables: {
        userId: 'me'
      }
    }
  ],
});
```

Let's see what the application looks like so far:

Figure 31.3 – The Todos MVC app with mutations

The input field for adding new todo items is just above the list of todo items. Now, let's look at the `TodoList` component, which is responsible for rendering the todo item list.

Rendering todo items

It's the job of the `TodoList` component to render the todo list items. When the `GET_USER` query takes place, the `TodoList` component needs to be able to render all the todo items. Let's take a look at the item list again, with several more todos added.

Figure 31.4 – The Todos MVC app with the filled list

Here's the `TodoList` component itself:

```
import React from 'react';
import { useMutation } from '@apollo/client';

import Todo from './Todo';
import { MARK_ALL_TODOS, GET_USER } from '../constants';

function TodoList({ user: { todos, totalCount,
  completedCount } }) {
  const [markAllTodos] = useMutation(MARK_ALL_TODOS, {
    refetchQueries: [{ query: GET_USER, variables:
      { userId: 'me' } }],
  });

  const handleMarkAllChange = () => {
    if (todos) {
      markAllTodos();
```

```
    }
  };

  return (
    <section className='main'>
      <input
        checked={totalCount === completedCount}
        className='toggle-all'
        onChange={handleMarkAllChange}
        type='checkbox'
      />

      <label htmlFor='toggle-all'>Mark all as
        complete</label>

      <ul className='todo-list'>
        {todos.map((todo) => (
          <Todo key={todo.id} todo={todo} />
        ))}
      </ul>
    </section>
  );
}

export default TodoList;
```

The `useMutation` Hook in this code block takes a mutation to check all todo items and returns a method that you can call from your React component. After doing so, it will refetch the query to get the user and their todos.

The relevant GraphQL query to get the data you need for this component is already executed in the `App` component. This component, therefore, doesn't need to send a query to the GraphQL backend itself and can render the todos that were passed to it. When you render the `App` component, you're passing it the `todo` data. Now, let's see what else we can do with the todos.

Completing todo items

The last piece of this application is rendering each todo item and providing the ability to change the status of the todo in the `Todo` component in `src/components/Todo.js`. Let's look at pieces of this code:

```
import classnames from 'classnames';
import { useMutation } from '@apollo/client';

import {
  CHANGE_TODO_STATUS,
  REMOVE_TODO,
  GET_USER,
} from '../constants';

function Todo({ todo }) {
  const [changeTodoStatus] =
    useMutation(CHANGE_TODO_STATUS, {
    refetchQueries: [{ query: GET_USER, variables:
      { userId: 'me' } }],
  });

  const [removeTodo] = useMutation(REMOVE_TODO, {
    refetchQueries: [{ query: GET_USER, variables:
      { userId: 'me' } }],
  });

  const handleCompleteChange = (e) => {
    const complete = e.currentTarget.checked;

    changeTodoStatus({ variables: { id: todo.id, complete }
      });
  };

  // ...
}
```

Here, you can see that this component is using two mutations, one to change the status of a todo item and another to remove a todo item. These mutations are passed to the `useMutation()` Hooks, which return a callback function to execute them. The callback functions are wrapped in different methods for reusability. The mutation to change the status of a todo can be found in `src/constants.js`:

```
import { gql } from '@apollo/client';

export const CHANGE_TODO_STATUS = gql'
  mutation ChangeTodoStatus($id: Int!, $complete: Boolean) {
    changeTodoStatus(id: $id, complete: $complete) {
      id
      complete
    }
  }
';
```

Based on the `id` type of the todo item, this mutation sends the request to the GraphQL backend to change the todo state. The GraphQL backend then talks to any services that are needed to make this happen. Then, it will refetch the `GET_USER` query to get the new list of todos, including the one you've just updated.

The actual component, that's making this possible is rendered by the `Todo` component, is a switch control, and the item text. When the user marks the todo as complete, the item text is styled as crossed off. The user can also uncheck items:

```
  return (
    <li
      className={classnames({
        completed: todo.complete,
        editing: isEditing,
      })}
    >
      <div className='view'>
        <input
          checked={todo.complete}
          className='toggle'
          onChange={handleCompleteChange}
          type='checkbox'
```

```
          />

          <label>{todo.text}</label>
          <button className='destroy'
            onClick={handleDestroyClick} />
        </div>
      </li>
    );
  };

  export default Todo;
```

That's all for the React implementation of the Todo app, but if you head over to the GitHub repository, you can see the full code source, including the example for a React Native implementation. In the native application, we've only used several queries and mutations, but the web version features more.

Summary

In this chapter, you implemented some specific GraphQL and Apollo Client ideas. Starting with the GraphQL schema, you learned how to declare the data that's used by the application and how these data types resolve to specific data sources, such as API endpoints. Then, you learned about bootstrapping GraphQL queries with Apollo Client in a React app. Next, you walked through the specifics of adding, changing, and listing todo items. The application itself uses the same schema as the web version of the Todo application, which makes things much easier when you're developing web and native React applications.

Well, that's a wrap for this book. We've gone over a lot of material together, and I hope that you've learned as much from reading it as I have from writing it. If there is one theme from this book that you should walk away with, it's that React is simply a rendering abstraction. As new rendering targets emerge, new React libraries will emerge as well. As developers think of novel ways to deal with state at scale, you'll see new techniques and libraries released. My hope is that you're now well prepared to work in this rapidly evolving React ecosystem.

Further reading

Refer to the following link for more information:

- Todo MVC: `https://todomvc.com/examples/react/`

Index

A

B

C

D

E

F

G

L

M

N

O

P

R

S

T

U

V

W

Hi!

We're Adam, Roy (@gethackteam on Twitter), and Mikhail, the authors of React and React Native Fourth Edition. We really hope you enjoyed reading this book and found it useful for increasing your productivity and efficiency in React.

It would really help us (and other potential readers!) if you could leave a review on Amazon sharing your thoughts on React and React Native Fourth Edition here.

Go to the link below or scan the QR code to leave your review:

`https://packt.link/r/1803231289`

Your review will help us to understand what's worked well in this book, and what could be improved upon for future editions, so it really is appreciated.

Best Wishes,

Mikhail Sakhniuk

Roy Derks

Packt.com

Subscribe to our online digital library for full access to over 7,000 books and videos, as well as industry leading tools to help you plan your personal development and advance your career. For more information, please visit our website.

Why subscribe?

- Spend less time learning and more time coding with practical eBooks and Videos from over 4,000 industry professionals
- Improve your learning with Skill Plans built especially for you
- Get a free eBook or video every month
- Fully searchable for easy access to vital information
- Copy and paste, print, and bookmark content

Did you know that Packt offers eBook versions of every book published, with PDF and ePub files available? You can upgrade to the eBook version at packt.com and as a print book customer, you are entitled to a discount on the eBook copy. Get in touch with us at customercare@packtpub.com for more details.

At www.packt.com, you can also read a collection of free technical articles, sign up for a range of free newsletters, and receive exclusive discounts and offers on Packt books and eBooks.

Other Books You May Enjoy

If you enjoyed this book, you may be interested in these other books by Packt:

React Projects

Roy Derks

ISBN: 978-1-80107-063-8

- Create a wide range of applications using various modern React tools and frameworks
- Discover how React Hooks modernize state management for React apps
- Develop web applications using styled and reusable React components
- Build test-driven React applications using Jest, React Testing Library, and Cypress
- Understand full-stack development using GraphQL, Apollo, and React
- Perform server-side rendering using React and Next.js
- Create animated games using React Native and Expo
- Design gestures and animations for a cross-platform game using React Native

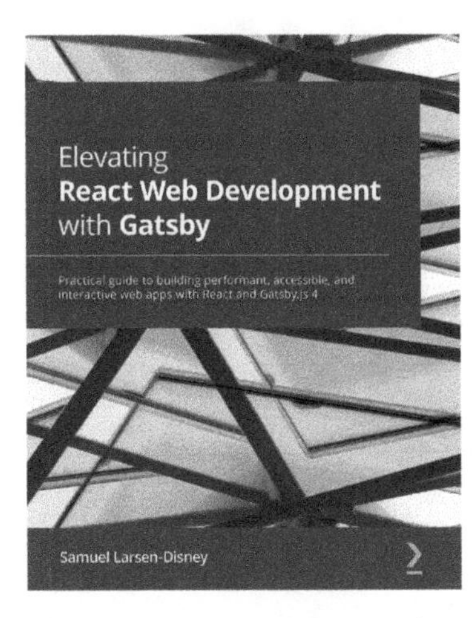

Elevating React Web Development with Gatsby

Samuel Larsen-Disney

ISBN: 978-1-80020-909-1

- Understand what GatsbyJS is, where it excels, and how to use it
- Structure and build a GatsbyJS site with confidence
- Elevate your site with an industry-standard approach to styling
- Configure your GatsbyJS projects with search engine optimization to improve their ranking
- Get to grips with advanced GatsbyJS concepts to create powerful and dynamic sites
- Supercharge your site with translations for a global audience
- Discover how to use third-party services that provide interactivity to users

Packt is searching for authors like you

If you're interested in becoming an author for Packt, please visit `authors.packtpub.com` and apply today. We have worked with thousands of developers and tech professionals, just like you, to help them share their insight with the global tech community. You can make a general application, apply for a specific hot topic that we are recruiting an author for, or submit your own idea.

CPSIA information can be obtained
at www.ICGtesting.com
Printed in the USA
LVHW022309170723
752627LV00005B/401